Nadia Nedjah, Luiza de Macedo Mourelle, Janusz Kacprzyk, Felipe M.G. França, and Alberto Ferreira de Souza (Eds.)

Intelligent Text Categorization and Clustering

Studies in Computational Intelligence, Volume 164

Editor-in-Chief

Prof. Janusz Kacprzyk
Systems Research Institute
Polish Academy of Sciences
ul. Newelska 6
01-447 Warsaw
Poland
E-mail: kacprzyk@ibspan.waw.pl

Further volumes of this series can be found on our homepage: springer.com

Vol. 142. George A. Tsihrintzis, Maria Virvou, Robert J. Howlett and Lakhmi C. Jain (Eds.)
New Directions in Intelligent Interactive Multimedia, 2008
ISBN 978-3-540-68126-7

Vol. 143. Uday K. Chakraborty (Ed.)
Advances in Differential Evolution, 2008
ISBN 978-3-540-68827-3

Vol. 144. Andreas Fink and Franz Rothlauf (Eds.)
Advances in Computational Intelligence in Transport, Logistics, and Supply Chain Management, 2008
ISBN 978-3-540-69024-5

Vol. 145. Mikhail Ju. Moshkov, Marcin Piliszczuk and Beata Zielosko
Partial Covers, Reducts and Decision Rules in Rough Sets, 2008
ISBN 978-3-540-69027-6

Vol. 146. Fatos Xhafa and Ajith Abraham (Eds.)
Metaheuristics for Scheduling in Distributed Computing Environments, 2008
ISBN 978-3-540-69260-7

Vol. 147. Oliver Kramer
Self-Adaptive Heuristics for Evolutionary Computation, 2008
ISBN 978-3-540-69280-5

Vol. 148. Philipp Limbourg
Dependability Modelling under Uncertainty, 2008
ISBN 978-3-540-69286-7

Vol. 149. Roger Lee (Ed.)
Software Engineering, Artificial Intelligence, Networking and Parallel/Distributed Computing, 2008
ISBN 978-3-540-70559-8

Vol. 150. Roger Lee (Ed.)
Software Engineering Research, Management and Applications, 2008
ISBN 978-3-540-70774-5

Vol. 151. Tomasz G. Smolinski, Mariofanna G. Milanova and Aboul-Ella Hassanien (Eds.)
Computational Intelligence in Biomedicine and Bioinformatics, 2008
ISBN 978-3-540-70776-9

Vol. 152. Jarosław Stepaniuk
Rough – Granular Computing in Knowledge Discovery and Data Mining, 2008
ISBN 978-3-540-70800-1

Vol. 153. Carlos Cotta and Jano van Hemert (Eds.)
Recent Advances in Evolutionary Computation for Combinatorial Optimization, 2008
ISBN 978-3-540-70806-3

Vol. 154. Oscar Castillo, Patricia Melin, Janusz Kacprzyk and Witold Pedrycz (Eds.)
Soft Computing for Hybrid Intelligent Systems, 2008
ISBN 978-3-540-70811-7

Vol. 155. Hamid R. Tizhoosh and M. Ventresca (Eds.)
Oppositional Concepts in Computational Intelligence, 2008
ISBN 978-3-540-70826-1

Vol. 156. Dawn E. Holmes and Lakhmi C. Jain (Eds.)
Innovations in Bayesian Networks, 2008
ISBN 978-3-540-85065-6

Vol. 157. Ying-ping Chen and Meng-Hiot Lim (Eds.)
Linkage in Evolutionary Computation, 2008
ISBN 978-3-540-85067-0

Vol. 158. Marina Gavrilova (Ed.)
Generalized Voronoi Diagram: A Geometry-Based Approach to Computational Intelligence, 2009
ISBN 978-3-540-85125-7

Vol. 159. Dimitri Plemenos and Georgios Miaoulis (Eds.)
Artificial Intelligence Techniques for Computer Graphics, 2009
ISBN 978-3-540-85127-1

Vol. 160. P. Rajasekaran and Vasantha Kalyani David
Pattern Recognition using Neural and Functional Networks, 2009
ISBN 978-3-540-85129-5

Vol. 161. Francisco Baptista Pereira and Jorge Tavares (Eds.)
Bio-inspired Algorithms for the Vehicle Routing Problem, 2009
ISBN 978-3-540-85151-6

Vol. 162. Costin Badica, Giuseppe Mangioni, Vincenza Carchiolo and Dumitru Dan Burdescu (Eds.)
Intelligent Distributed Computing, Systems and Applications, 2008
ISBN 978-3-540-85256-8

Vol. 163. Pawel Delimata, Mikhail Ju. Moshkov, Andrzej Skowron and Zbigniew Suraj
Inhibitory Rules in Data Analysis, 2009
ISBN 978-3-540-85637-5

Vol. 164. Nadia Nedjah, Luiza de Macedo Mourelle, Janusz Kacprzyk, Felipe M.G. França and Alberto Ferreira de Souza (Eds.)
Intelligent Text Categorization and Clustering, 2009
ISBN 978-3-540-85643-6

Nadia Nedjah
Luiza de Macedo Mourelle
Janusz Kacprzyk
Felipe M.G. França
Alberto Ferreira de Souza
(Eds.)

Intelligent Text Categorization and Clustering

Springer

Nadia Nedjah
Universidade do Estado do Rio de Janeiro
Faculdade de Engenharia
sala 5022-D
Rua São Francisco Xavier 524
20550-900, MARACANÃ-RJ, Brazil
Email: nadia@eng.uerj.br

Luiza de Macedo Mourelle
Universidade do Estado do Rio de Janeiro
Faculdade de Engenharia
sala 5022-D
Rua São Francisco Xavier 524
20550-900, MARACANÃ-RJ, Brazil
Email: ldmm@eng.uerj.br

Prof. Janusz Kacprzyk
Systems Research Institute
Polish Academy of Sciences
Ul. Newelska 6
01-447 Warsaw, Poland
Email: kacprzyk@ibspan.waw.pl

Prof. Felipe M.G. França
Programa de Engenharia de Sistemas e
ComputaçÃo
COPPE/UFRJ
Caixa Postal 68511
21941-972 - Rio de Janeiro - RJ, Brazil
Email: felipe@cos.ufrj.br

Alberto Ferreira de Souza
Pró-Reitor Planejamento e Desenvolvimento
Institucional - UFES
Av. Dante Micheline, 2431, APT 803
29066-430 - Vitória- ES, Brazil
Email: alberto@inf.ufes.br

ISBN 978-3-642-09929-8 e-ISBN 978-3-540-85644-3

DOI 10.1007/978-3-540-85644-3

Studies in Computational Intelligence ISSN 1860949X

Printed in acid-free paper

9 8 7 6 5 4 3 2 1

springer.com

Preface

Automatic Text Categorization and Clustering are becoming more and more important as the amount of text in electronic format grows and the access to it becomes more necessary and widespread. Well known applications are *spam* filtering and web search, but a large number of everyday uses exist (intelligent web search, data mining, law enforcement, etc.) Currently, researchers are employing many intelligent techniques for text categorization and clustering, ranging from support vector machines and neural networks to Bayesian inference and algebraic methods, such as Latent Semantic Indexing.

This volume offers a wide spectrum of research work developed for intelligent text categorization and clustering. In the following, we give a brief introduction of the chapters that are included in this book.

In Chapter 1, the authors present the use of attribute selection techniques to define a subset of genes related to specific characteristics such as cancer arising. Through combination of search methods and evaluation procedures, the authors show that the data mining algorithm speeds up, mining performance such as predictive accuracy is improved and the comprehensibility of the results becomes easier in most of the combinations. The authors obtained best results with wrapper approaches and sequential search.

In Chapter 2, the authors propose a new preprocessing technique for online handwriting. The approach is to first remove the hooks of the strokes by using changed-angle threshold with length threshold, then filter the noise by using a smoothing technique, which is the combination of the Cubic Spline and the equal-interpolation methods. Then, the handwriting is normalised.

In Chapter 3, the authors explore clustering of unstructured document collection. They explore a simple procedure that not only considerably reduces the dimension of the feature space and hence the processing time, but also produces clustering performance comparable or even better when confronted with the full set of terms.

In Chapter 4, the authors investigate the application of query expansion technique to improve cross-language information retrieval in English and Thai as well as the potential to apply the technique to other intelligent systems such as tutoring systems. As a method of evaluation of query expansion, they attempt to find out whether the expanded terms are useful for the search.

In Chapter 5, the authors provide a fuzzy partition and a prototype for each cluster by optimizing a criterion dependent on the dissimilarity function chosen. They include experiments involving benchmark data sets and carried out in order to compare the accuracy of each function. In order to analyse the results, they apply an external criterion that compares different partitions of a same data set.

In Chapter 6, the authors describe a system for cluster analysis of hypertext documents based on genetic algorithms. The system's effectiveness in getting groups with similar documents is evidenced by the experimental results.

We are very much grateful to the authors of this volume and to the reviewers for their tremendous service by critically reviewing the chapters. The editors would also like to thank Prof. Janusz Kacprzyk, the editor-in-chief of the Studies in Computational Intelligence Book Series and Dr. Thomas Ditzinger from Springer-Verlag, Germany for their editorial assistance and excellent collaboration to produce this scientific work. We hope that the reader will share our excitement on this volume and will find it useful.

May 2008

Nadia Nedjah
Luiza M. Mourelle
Felipe M.G. França
Janusz Kacprzyk
Alberto F. de Souza

Contents

List of Figures

List of Tables

1

Gene Selection from Microarray Data

Helyane Bronoski Borges[1] and Julio Cesar Nievola[2]

[1] UTFPR - Universidade Tecnológica Federal do Paraná
helyane@utfpr.edu.br
[2] PPGIa - Pontificia Universidade Católica do Paraná (PUCPR)
nievola@ppgia.pucpr.br

This chapter presents the use of attribute selection techniques to define a subset of genes related to specific characteristics such as cancer arising. Through combination of search methods and evaluation procedures, it is showed that the data mining algorithm speeds up, mining performance such as predictive accuracy is improved and the comprehensibility of the results becomes easier in most of the combinations. Best results were achieved with wrapper approaches and sequential search.

1.1 Introduction

The automatic knowledge discovery in databases has been applied in the most different areas as text mining (classification and categorization) and in gene expression database obtained by the microarray approach. The analysis of these kind of data type presents common particularities comparing with other databases. For instance, the number of samples is small (in the order of dozens or hundreds of samples), and with a huge number of attributes, where each one correspond to each one of the words in the set of texts or the genes that have their pattern expressed in the samples (both typically in the order of dozens of thousands of elements).

Text classification is a very important task, being used in many applications. It is done by assigning labels to documents based on a set of previous labelled documents. In this case, the first step is to represented the documents as vectors in a high-dimension vector space, where each dimension corresponds to the value of a feature, specifically, a term. The classifiers use these vectors to classify the documents to the right categories. In order to do so, it is important to decide the relevance of every term, known as term weighting; this is a critical decision to the performance of the classification result. Before doing term weighting, terms should be a selection of the most important terms, that means, the ones that would contribute the most for the classification. Inverse document frequency (IDF) is one of the most popular methods for this task; however, in some situations, such as supervised learning for text categorization, it doesn't weight terms properly, because it neglects the category information and assumes that a term that occurs in smaller set of documents should get a higher weight. There

N. Nedjah et al. (Eds.): Intelligent Text Categorization and Clustering, SCI 164, pp. 1–23.
springerlink.com

have been several term weighting schemes that consider the category information. Evaluation functions, such as IG (Information Gain), GR (Gain Ratio), MI (Mutual Information) and CHI (chi-square) [8], [12], [44] are used in the term weighting step [13]. There is also a powerful term weighting scheme called ConfWeight (Confidence Weight) [40], that considers the category information, and show a good performance.

Text clustering is another task that is receiving considerable attention in areas such as text mining and information retrieval. Clustering techniques also start with an unstructured document collection based on vector-space model (usually known as bag-of-words model [36]). The central idea is to use the terms as features and then describe each document by a vector that represents the frequency occurrence of each term in the document. The set of documents is then described by the so-called document-to-term matrix whose ij element indicates the frequency of term j in the document i. Therefore, this matrix has ND rows and NT columns where ND and NT are respectively the total numbers of documents and terms. This matrix tends to be huge basically due to the exceeding number of terms.

In both cases, the use of techniques to diminish the number of features (terms) regarding the number of samples leads to models that presents less problems regarding their generalization, that gets generally better quality indexes and also are easier to understand than the models built using the full set of terms. Another area that has the same challenge due to its huge number of attributes relative to the samples is the classification or clustering of data in molecular biology, specifically dealing with micro-arrays experiments. In the later, techniques originated from the data mining field are in use for some time and have demonstrated their viability and conducted to an improvement of the practical results obtained in the field.

The data mining is one of the main steps in the knowledge discovery process. Among the tasks developed in this step are the clustering and classification.

The clustering and classification tasks are used very often used in Biology area to determine groups of data that have similar behavior and to classify new samples according to them. The classification approach is more often used as a second stage after the cluster discovery (particularly in the clinical analysis with diagnosis and prognostic purposes).

Most of the time when studying gene expression data, data with unknown characteristics are manipulated, which can be redundant and even, in certain cases, irrelevant. Since the amount of attributes used has great influence in the learning algorithms, the identification of those attributes really important for an application becomes a factor of extreme importance to be successful. The exclusion of irrelevant or redundant attributes almost always improves the quality of the obtained results. Considering the classification task, Yang and Honavar [43] affirm that the choice of the attribute subset affects, among other factors, the precision of a classifying system, its cost, and the time and the amount of necessary examples for the learning task. This activity encompasses a correlated classification task known as attribute subset selection.

The attribute subset selection can be defined, according to Liu and Motoda [23], as the process that chooses the best attribute subset according to a certain criterion.

1.2 Related Works

Microarray data has been around for a while, but only recently computational techniques started to be used to deal with it to produce useful knowledge [32], [28], [11]. Initially, statistical techniques were used alone to envisage relationships among the genes and its links to diseases [2], [21]. Following that, the Machine Learning community started to use its own techniques in order to determine how far one can extract information from experiments obtained from microarrays [35]. Data Mining practitioners see this kind of data as challenging due to its very nature, that means, only a few examples (in the order of dozens to hundreds) are available, despite the number of attributes (genes) is huge (in the order of thousands to tens of thousands) [2] [31]. In this case, the three main paradigms in the area were used: classification, clustering and visualization [20]. For classification, various algorithms were used [18] [27], especially SVM [9] [10]. The use of the clustering approaches [41] [37] ranges from associative memories [3] to bi-clustering techniques [38]. Some tools were also developed to integrate validation techniques [4]. One of the most researched topics is the selection of the relevant genes (from a microarray data) linked to specific characteristics [42] [19] [7].

Until some years ago most of the studies accomplished in gene expression databases involved the application of non-supervised learning with the use of clustering techniques [1], [15].

An example of class discovery using clustering techniques is the one of Alizadeh and her colleagues [1]. In that work, the authors have identified two distinct forms of Diffuse Large B Cell Lymphoma.

The dataset employed consists in 47 samples, 24 of them belonging to the germinative center of B cell group, while 23 belong to the activation B cell group. Each sample is represented by 4026 genes, all of them with its numeric value. The 4027th value is the goal attribute. The goal is to identify to which class each sample is related: germinal or activated.

Rosenwald and her colleagues used the clustering approach to classify the patient with Diffuse Large B Cell Lymphoma survival condition, after the chemotherapy. The database is formed by 240 patient's samples and 7399 genes [34].

The supervised approach represents a powerful alternative that can be applied when previous information on the class of the genes exists. For instance, the techniques of Supervised Machine Learning can be used for the prediction and classification tasks.

An example of supervised learning application can be found in [39]. The authors used the vote algorithm to classify the two types of Lymphoma: Diffuse Large B Cell Lymphoma (DLBCL) and Follicular Lymphoma (FL) analyzing 6817 genes obtained from 77 patients (58 belonging to the group of DLBCL and

19 belonging to the group of FL). With that technique they obtained 91% of success.

The success of the classification to distinguish the two tumors types motivated the authors to evaluate the patients' survival conduction with DLBCL (58 patients). The set was formed by 32 samples belonging to the cured patients' group and 26 samples belonging to the no cured patients' group. For the classification of that set task they used two algorithms: SVM (Support Vector Machines) and k-NN (k-Nearest Neighbor) and they obtained of 72% and 68% a rate of success respectively.

As already commented, the microarray data presents challenges to the machine learning algorithms, which are usually developed to work with a high number of samples (examples) and with relatively few attributes or features.

Borges and Nievola [5], [6], using the same database used by Alizadeh and colleagues in their experiments, applied attribute selection algorithms looking for improvement in the results of the classification algorithms.

To apply the attribute selection methods in that work, the authors divided the work in two phases: the search for attributes subsets and the evaluation of the found subsets.

Two methods of search were used: the sequential search and the random search. For subsets evaluation it was used two main approaches: the filter and the wrapper approaches. As evaluation measures of the filter approach the dependence and the consistency measure were used. The wrapper approach used the data mining algorithm itself for the subsets evaluation, where the algorithms Naïve Bayes, Bayesian Networks, C4.5, Decision Table and k-NN (for the k=1, k=3, k=5 and k=7) were chosen.

In another work Borges and Nievola [7], applied the same attribute selection algorithms in the DLBCL-Tumor and BLBCL-Outcome database. In these experiments, it has been observed that the use of attribute selection, despite the huge number of attributes and small number of samples, has lead to a better performance of the classifiers in almost all cases and schemes on both datasets. Random search is not advisable, considering that its results are hard predictable. The use of sequential search, on the other hand, consistently achieved improvements in the classification.

The results also indicate that the sequential forward search algorithm is to be used, considering not only its best computational results, but also that it matches biological knowledge saying that in general few genes interact in order to give raise to a characteristic, in this case the cancer.

1.3 Knowledge Discovery Database - KDD

Most of the times, when working with genetic expression data, the data characteristics are unknown, since they can be redundant and sometimes even irrelevant to the question under consideration. In order to diminish the associated problems when these characteristics are present, an initial step of preprocessing is recommended, to achieve better data quality and, consequently, better

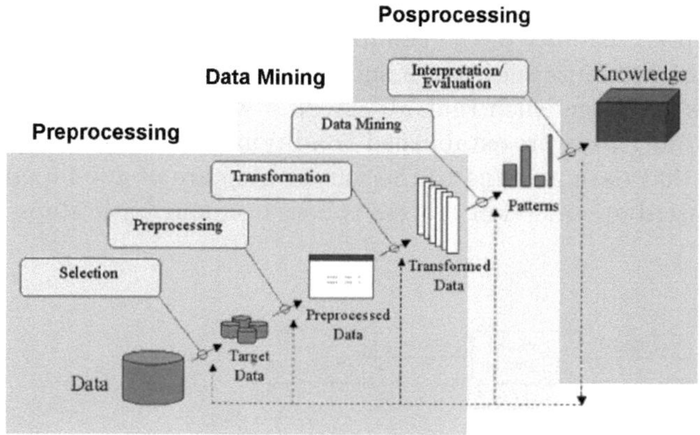

Fig. 1.1. KDD process phases [14]

performance of the data mining algorithm [22]. With this intention, the classification task was preceded by the application of some attribute selection schemes and their result in the whole process evaluated.

Knowledge Discovery Database is the discovering implicit and useful information process in databases [14]. The KDD process is defined by a stages group, involving from the understanding of the domain of the application to the interpretation and consolidation of the results.

Fayyad et al. (1996) represent a typical KDD process in the Figure 1.1 divided in three great phases as the preprocessing, that contemplates the selection, processing and transformation of the data, the data mining discovering patterns, and the posprocessing that contemplates the patterns and knowledge discovered.

The KDD involves several complex stages that should be executed correctly for that the established objectives and the success of the application are reached. The process is interactive, with a lot of decisions to be taken, and iterative, could possess loops among any of the stages, not existing an order or only sequence during the course of the process [14].

1.3.1 Attribute Selection

One important step in the KDD process is data preprocessing. This stage is responsible for the consolidation of the relevant information to the mining algorithm, trying to reduce the problem's complexity. Among the various substeps in the data mining preprocessing, one should pay attention to the attribute selection.

Attribute selection is one of the most important data preprocessing techniques and is often used. It reduces the number of characteristics to be used, removes irrelevant, redundant and noisy data, and brings immediate effects to the application at hand: it improves data mining algorithm's speed, improves precision and comprehensibility of the results.

Through the attribute selection a subset of M attributes out of the N original attributes is chosen, such that $M \leq N$, in a way the characteristics space is reduced according to a pre-established condition [23]. Attribute selection tries to guarantee that data arriving to the mining step are of good quality.

A typical attribute selection process is based on four basic steps, as shown in Figure 1.2.

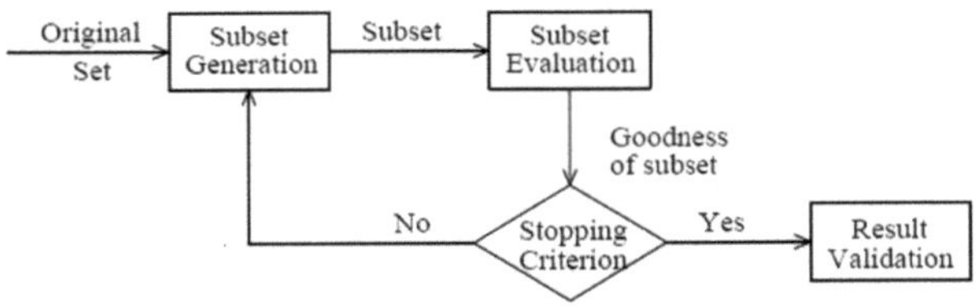

Fig. 1.2. Attribute Selection steps [23]

The subset generation is a search procedure that creates candidates attributes subsets based on a search strategy to be evaluated. Each generated subset is evaluated and compared with the previous best one, according to an evaluation criterion. If the new subset is better than the old one, it replaces that. If the new subset is worst than the old one, it is discarded and the old subset remains as the better choice. The generation and evaluation process is repeated until some stopping criterion is reached. When it happens, the best subset found needs to be validated through some a priori knowledge or different tests by a real or synthetic dataset [26].

This process's nature is dependent on two basic aspects. First, one should decide the search starting point (or points), which will determine the search direction. The search can start with an empty set and steadily adding attributes to the existing set (forward search), or it can start with a full set of attributes and removing one attribute at each iteration (backward search). Another possibility is to start with a predefined set of attributes and adding/removing attributes at each step. When the search starts with a randomly selected subset of attributes it can avoid the trap of a local minimum. The second decision to be made is regarding the search strategy. For a dataset with N attributes, there are 2^N possible subsets. This search space grows exponentially, making it prohibitive, even for a moderate value of N. According to this line of reasoning the search algorithms can be divided in three main groups: exponential, sequential and random algorithms.

Exponential algorithms, such as the exhaustive search, use all the possible combinations of the attributes before returning the resulting attribute subset. They are computationally infeasible, due to the rapidly (exponential) growth of the running time as the number of available attributes arises [23].

Sequential algorithms, such as the forward sequential selection and the backward sequential selection, are very efficient in solving many problems of attribute selection. Their main disadvantage is that they don't take into account attribute interaction.

Forward sequential selection starts the quest for the best subset attribute with an empty set of attributes. Initially, attribute subsets with only one attribute are evaluated, and the best attribute A^* is selected. This attribute A^* is then combined with all other attributes (in pairs), and the best attribute subset is selected. The search continues on, always adding one attribute at a time to the best attribute subset already selected, until no improvement on the quality of the attribute subset is possible.

Backward sequential selection, contrarily to forward sequential selection, starts the search for the subset of optimal attributes with a set encompassing all attributes, and at each iteration one attribute is removed from the actual solution, until no further improvement on the quality of the solution can be made.

Each new subset that is created needs to be evaluated using the evaluation criterion. The quality of a subset can be computed according to a certain criterion (for instance, a selected optimum subset through the use of one criterion could not be optimal according to another criterion). In a general way, the evaluation criteria can be categorized in two groups, based on whether it is dependent or not on the mining algorithm that will be applied to the resulting optimal attribute subset.

The filter approach (Figure 1.3) characterizes the independent criterion [22]. It tries to evaluate an attribute (or a subset of attributes) exploring intrinsic characteristics of the training data, without any commitment to the mining algorithm.

The most used independent criterion are: distance measures, information measures, dependence measures and consistency measures.

The distance measure is also known as separability, divergence, or discrimination measure. For a two-class problem, an attribute X is preferred to another attribute Y if X induces a greater difference between the two-class conditional probabilities than Y; if the difference is zero, then X and Y are indistinguishable. An example is the Euclidean distance measure [26].

The information measure typically determine the information gain from an attribute. The information gain from an attribute X is defined as the difference between the prior uncertainty and expected posterior uncertainty using X. Attribute X is preferred to attribute Y if the information gain from attribute X is greater than that from attribute Y(e.g., entropy measure) [26].

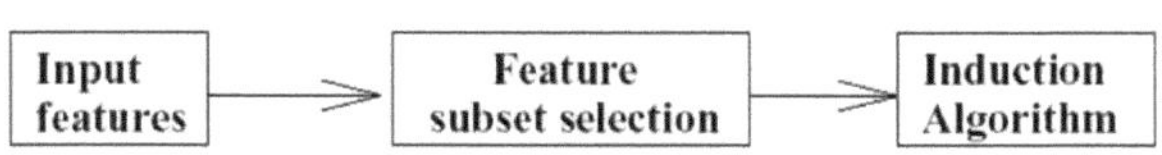

Fig. 1.3. The filter approach to attribute subset selection [22]

The dependence measure is also known as correlation or similarity measure. This measure quantifies how much two attributes are associated, implying that the knowledge of one attribute allows the prediction of the second attribute. In attribute selection context, the best evaluated attribute is the one that best predicts the class (the value of the goal attribute).

It measures the ability to predict the value of one variable given the value of another one. For classification purposes, the better attribute is the one that better predicts the class. An attribute X is better than another attribute Y if the association between attribute X and class C is higher than the association between Y and C.

One of the algorithms that use this measure is CFS (Correlation-based Feature Selection) [16]. This algorithm evaluates the importance of an attribute subset based on the individual predictive ability of each attribute and the correlation degree among them. The merit equation in this case is given by:

$$Merit_s = \frac{k\bar{r}_{cf}}{\sqrt{k + k(k-1)\bar{r}_{ff}}} \tag{1.1}$$

In equation (1), *Merit* is the heuristic of the attribute subset S containing k attributes, is the average attribute-class correlation, and is the average attribute-attribute inter-correlation [16].

The difference between CFS and other filter algorithms is that while the general filter algorithms supply an independent score for each attribute, CFS presents a heuristic reward of the attribute subset and informs the best subset found.

The consistency measure is characteristically different from the above measures because of their heavy reliance on the class information and the use of the Min-Features bias in selecting a subset of features [23]. These measures attempt to find a minimum number of features that separate classes as consistently as the full set of features can. An inconsistency is defined as two instances having the same feature values but different class labels. As example, an inconsistency in X' and S is defined by two instances in S being equal when considering only the features in X' and that belong to different classes. The aim is, thus, to find the minimum subset of features leading to zero inconsistencies [25]. The inconsistency count of an instance is defined as:

$$IC_{x'}(A) = X'(A) - max_k X'_k(A) \tag{1.2}$$

In Equation (2), $X'(A)$ is the number of instances in S equal to A using only the features in X' and $Xk'(A)$ is the number of instances in S of the class k equal to A using only the features in X'. The inconsistency rate of a feature subset in a sample S is then given by:

$$IR(X') = \frac{\sum_{A \varepsilon S} IC_{X'}(A)}{|S|} \tag{1.3}$$

This is a monotonic measure in the sense: $X_1 \subset X_2 \Rightarrow IR(X_1) \geq IR(X_2)$

A possible evaluation measure is then:

$$J(X') = \frac{1}{IR(X') + 1} \tag{1.4}$$

This measure is in the range [0,1] and can be evaluated in $O(|S|)$ time using a hash table.

The dependent criterion, characterizes the wrapper approach shown in Figure 1.4 [22]. It requires a predefined mining algorithm within the attribute selection and uses its performance when applied in the selected subset to evaluate the quality of the selected attributes.

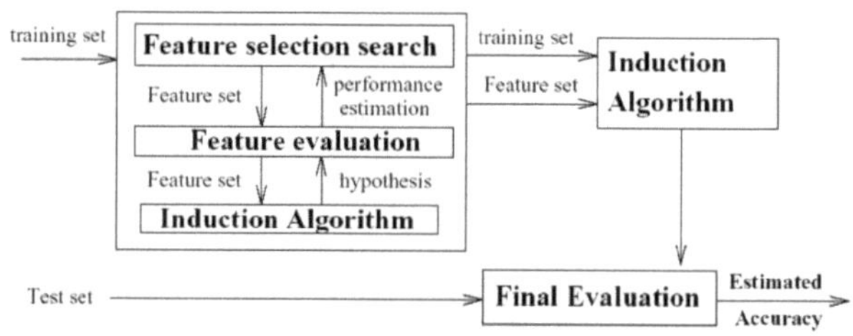

Fig. 1.4. The wrapper approach to attribute subset selection [22]

The stopping criterion establishes when the attribute selection process should be stopped. It can be done when the search is over, or when the goal is reached, where the goal can be an specific situation (maximum number of characteristics or maximum number of iterations), or when a sufficiently good subset is found (for instance, a subset could be sufficiently good if the classification error rate is under some threshold for a given task).

1.4 Experiments

Some were made experiments in microarray databases with objective of the attribute reduction or to aid of the knowledge discovery. This section shown the methodology used in the experiments and the results found.

1.4.1 Methodology

This experimenters was developed by following the stages of the knowledge discovery process based on the three main phases: preprocessing, data mining and pos-processing. For the execution of some experiments Weka software was used.

Database description

For the experiments two databases selected and identified were DLBCL-Outcome [39] and DLBCL-NIH [34]. They refer to the lymphoma study already cited aiming at the classification of patient's survival condition after a treatment period.

The DLBCL-NIH database is formed by 240 patient samples (102 alive and 138 no alive) and 7399 attributes (genes). The objective with this database is to analyze the patient's survival condition with Diffuse Lymphoma of Large Cell B after the chemotherapy.

The DLBCL-Outcome database is composed by 58 samples, 32 belonging to the cured patient's group and 26 to the patient's group no cured, using 7129 attributes. This database presents the result of the DLBCL patient's treatment after 5 years. The data in this base identify which patients had success in the treatment and were cured and the ones with no cure

Preprocessing

The preprocessing phase the attribute selection was used to reduce the dimensionality of the data. Two kinds of search were used: sequential and random. Used evaluation measures were based on filter and wrapper approach. For the filter approach dependence and consistency measures were chosen.

For the wrapper approach, as it uses the mining algorithm for the subset evaluation, four classification algorithms were used: Naïve Bayes, induction decision tree executed by the algorithm C4.5, SVM and k-NN. All those algorithms were run considering their parameters configured with the standard values (standard values in Weka environment). For the SVM classifier the parameters used were: with value same to 1.0 and epsilon with value 1.0ϵ-12. For the k-NN classifier k=1, k=3, k=5 and k=7 values were used.

The sequential search applied in the experiments used a forward greedy search beginning with any attribute. The search stops when the attribute addition results in the reduction of the evaluation; in other words, the rate of misclassification increases.

The random search used in these experiments uses a simple genetic algorithm. The evaluation solution found (fitness function or aptitude function) is based on the number of instances correctly classified. The initial population was created based on a uniform random distribution. The group of the parameters was established a priori, with the size of the population and the number of generations being defined as 20, and the crossover and mutation probabilities defined as 0.6 and 0.033, respectively.

Table 1.1 shows the combinations of the search method and the successor's generation chosen for the experiments. The evaluation criterion chosen for the trials to are presented in Table 1.2.

Table 1.1. Search methods and successor's generation

Search Method	Successor's generation
Sequential	Forward
Random	Random

Table 1.2. Subset evaluation

Evaluation measures
Dependence
Consistency
Wrapper (with Naïve Bayes)
Wrapper (with C4.5)
Wrapper (with SVM)
Wrapper (with k-NN, k=1)
Wrapper (with k-NN, k=3)
Wrapper (with k-NN, k=5)
Wrapper (with k-NN, k=7)

Regarding the wrapper approach, the following learning algorithms were used: Naïve Bayes, C4.5, SVM and k-NN (using four different values for k, k=1, k=3, k=5 e k=7). When doing the classification of each selected attribute subset, the classifier used to do so was the same as in the attribute selection step.

In this work the strategy adopted was to stop the process when the subsequent addition (or deletion) of any attribute doesn't provide an improvement in the quality measure.

Data Mining

The objective of the work is the attribute reduction in microarray databases using the attribute selection. However, it is important to remind that, if the dimensionality reduction is excessive, the classifier can have its discrimination power reduced. Therefore, it is important to compare the variation of the classifier behavior regarding the database with all attributes, so that it is possible to estimate the ideal dimensionality of the dataset, for a defined classifier based on these results.

Therefore, two databases (with all of the attributes) were submitted to 4 classifiers: Naïve Bayes, C4.5, SVM and k-NN (for k=1, k=3, k=5 and k=7) as a way to compare the rate of success of the classifier using all attributes in relation to its behavior when using the subsets generated by reduction dimensionality methods. The indicated algorithms were chosen because they belong to different learning paradigms.

Posprocessing

The results were appraised through estimate of average of success of each classifier using stratified cross validation with 10 partitions (10 fold cross validation), which consists firstly, in dividing random the original database in same or approximately same k partitions, so much in the example number, as in the distribution of the classes. Like this, each partition has a same class or approximately equal distribution to the classes' distribution of the complete database. For that procedure it is selected k-1 partitions to form the training database and, soon afterwards, the amount the instance correctly classification is computed for the partition that was left out. This step repeats until that all of the k partitions, k-1 are used times as training database and once as test database. At the end, the performance is obtained by the arithmetic average obtained (in the test data) for the k partitions. Besides, for all the results were made statistical analyses to evaluate the significance of the results through the hypothesis test with p-value same to 0.05, which it establishes the tolerated mistake and, at the same time, it defines the null hypothesis rejection area.

The statistics significant term used in this work refers to the result of a algorithms' comparison using the statistics of the test-t pared, where the fundamental objective is to evaluate the behavior of the differences observed in each element [30], [46].

1.4.2 Results and Discussion

The two chosen databases were submitted to the 4 classifiers. In table 1.3 the results of the classification algorithms for each database are shown. The annotation "*" indicates that a result is statistically worse than the result of the standard algorithm (Naïve Bayes) and the result in bold indicate that the result is statistically better compared with the standard algorithm.

Table 1.3. Results of database classification with all the attributes

Algorithms	DLBCL-Outcome	DLBCL-NIH
Naïve Bayes	42,0 ± 24,8	59,6± 11,9
C4.5	53,3 ± 11,5	52,1 ± 9,9
SVM	54,3 ± 20,7	63,7 ± 11,5
1-NN	45,7 ± 24,6	51,2± 10,2
3-NN	38,7 ± 24,6	49,2 ± 11,6
5-NN	47,7 ± 23,8	50,8 ± 89,8
7-NN	53,0 ± 24,5	50,0 ± 7,7

No Attribute Selection

Initially the classification algorithms were executed on the original data for future comparisons with the attribute selection algorithms.

Figures 1.5 and 1.6 shows the results when applying the classifiers on the data using all attributes in the DLBCL-Outcome Database and DLBCL-NIH Database. The bar indicates the average classification rate for each classifier and the line represents the inferior and superior limit (the standard deviation). The figures shows that both databases had relatively low results. Analyzing the results of all classifiers one sees that all results are low. A possible cause of this

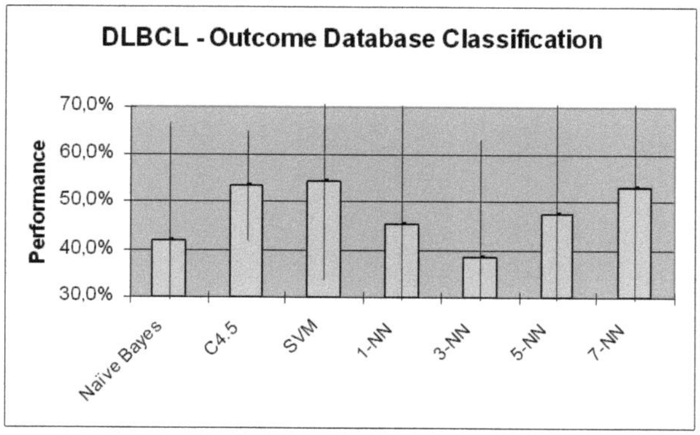

Fig. 1.5. Classification rate for each classifier with all attributes in the DLBCL-Outcome Database

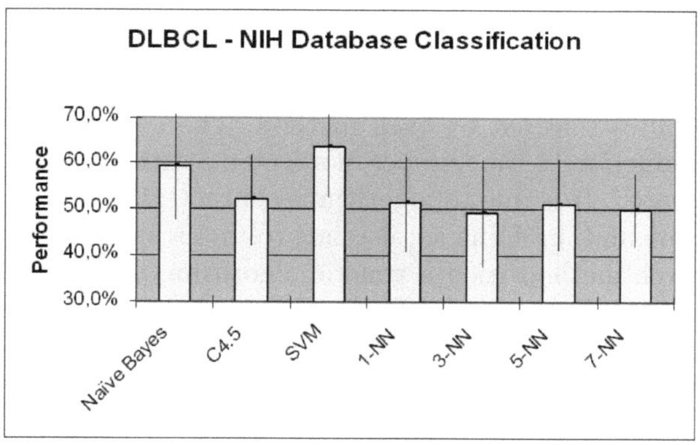

Fig. 1.6. Classification rate for each classifier with all attributes in the DLBCL-NIH Database

low classification rate can be due to the data type that is being used. Using these bases the result is based only on the interaction of some genes. One could argue (in a logical way) that the evolution of the treatment is also dependent on some other factors like the type of treatment, specific characteristics of the disease, the patient's characteristics (as age, sex, etc.). These information, however, are not available in the base and, therefore, not used.

Attribute Selection

Two approaches were used for the attribute selection: filter and wrapper. In filter approach the dependence and consistency evaluation measures were tested. In the case of the wrapper approach the following mining algorithms were selected and used as evaluation measures: Naïve Bayes W(NB), C4.5 W(C4.5), SVM

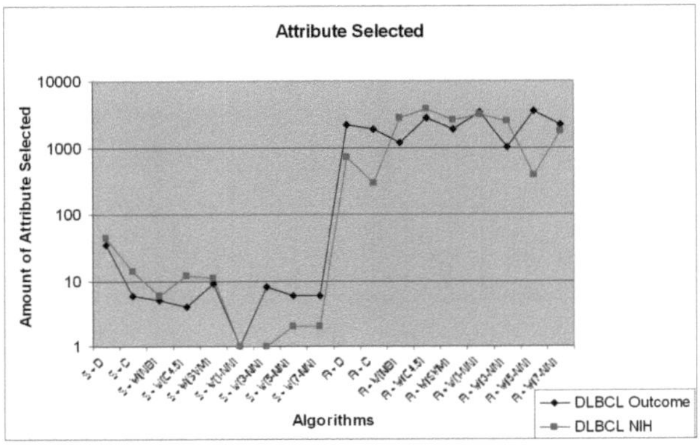

Fig. 1.7. Amount of Attributes Selected (observe that it is in logarithmic scale)

W(SVM) and the k-NN W(1-NN), W(3-NN), W(5-NN) and W(7-NN). As search criterion the sequential search (S) and the random search (R) were used.

It is possible to observe in the Figure 1.7 that there is great variation in the number of attributes selected by each method. The sequential search method caused a great reduction in the number of selected attributes compared with the use of random search. It is due to algorithms characteristics, because it adds an attribute at a time and, in doing so, it tends to find a smaller attributes subset. The random search method used a genetic algorithm, which selected a greater number of attributes. It happened due to the random choice of the attributes, that doesn't privilege a subset of a specific size.

Has observed that those two databases had relatively low results, however analyzing the results of those whole classifiers all the results equivalent statistic is considered. A possible cause of the drop performance can be due to the data type that if is analyzing. In those two bases the result is not dependent only of the interaction of some genes but of the treatment disease type, the patient's characteristics (as age, sex, etc.) among other factors however those data are not available in the database.

Filter Approach

For each attribute subset generated by the attribute selection algorithm classification algorithms were applied.

In Table 1.4 the results of the use of filter approach on each attribute subset of the DLBCL-Outcome database are presented. Only two of those results are considered worse, when compared with the Naïve Bayes algorithm: the results obtained with the C4.5 algorithm and with the algorithm 7-NN. Both results were worse in the execution of the sequential search method with the dependence evaluation measure.

In Table 1.5 one can observe that the k-NN algorithm presented the worst results. The k-NN algorithm with k=1 presented worse results in two methods

Table 1.4. Result of the classification of the attributes subsets of the DLBCL-Outcome database after the attribute selection using the filter approach

Algorithms	S + D	S + C	R + D	R + C
Naïve Bayes	84,0 ± 13,6	70,3 ± 25,1	35,6 ± 26,1	44,0 ± 23,0
C4.5	68,3 ± 17,2*	68,3 ± 15,3	46,3 ± 20,2	63,6 ± 20,7
SVM	77,3 ± 14,4	70,3 ± 20,6	59,3 ± 22,5	57,7 ± 21,1
1-NN	70,0 ± 28,5	77,0 ± 17,1	44,0 ± 27,2	44,0 ± 27,2
3-NN	73,0 ± 27,6	75,6 ± 12,2	40,3 ± 28,2	47,6 ± 23,8
5-NN	73,7 ± 20,4	65,6 ± 15,5	46,0 ± 24,2	49,3 ± 21,9
7-NN	73,6 ± 17,1*	69,0 ± 13,1	56,3 ± 27,8	51,3 ± 22,6

Table 1.5. Classification result of the attributes subsets of the DLBCL-NIH database after the attribute selection using the filter approach

Algorithms	S + D	S + C	R + D	R + C
Naïve Bayes	2,9 ± 10,6	66,2 ± 8,2	61,6 ± 11,4	60,4 ± 9,05
C4.5	66,2 ± 8,8	66,6 ± 6,8	57,5 ± 9,9	50,4 ± 10,6
SVM	65,4 ± 8,8	61,6 ± 11,9	56,3 ± 7,6	66,2 ± 12,0
1-NN	55,8 ± 8,15*	53,7 ± 11,2*	59,2 ± 7,8	53,7 ± 8,88
3-NN	57,1 ± 8,8	57,1 ± 6,5*	49,6 ± 9,1	47,9 ± 11,3*
5-NN	57,1 ± 7,6*	58,7 ± 6,3	50,8 ± 10,9	52,5 ± 11,9
7-NN	57,5 ± 7,1*	57,1 ± 6,2	49,6 ± 9,3*	52,5 ± 6,3

Table 1.6. Classification results of the attribute subsets of the DLBCL-Outcome database after the attribute selection using the wrapper approach

Algorithms	Sequential	Random
Naïve Bayes	89,7 ± 11,9	52,7 ± 21,7
C4.5	90,7 ± 13,7	48,0 ± 19,7
SVM	94,0 ± 13,5	59,6 ± 21,7
1-NN	77,6 ± 11,0*	52,3 ± 24,9
3-NN	89,6 ± 11,9	65,3 ± 19,4
5-NN	93,3 ± 11,6	51,0 ± 19,7
7-NN	86,0 ± 13,4	60,0 ± 21,6

of attribute selection, both in the execution of the sequential search method together with the two filter approach evaluation measures. The 3-NN algorithm presented worse results together in the sequential search method with the evaluation measure and in the method of random search with the consistency evaluation measure. In the 5-NN algorithm the worst results were in the execution of the sequential search method with the dependency evaluation measure and in the random search method with the dependency evaluation measure. Finally,

Table 1.7. Classification result of the attributes subsets of the DLBCL-NIH database after the attribute selection using the wrapper approach

Algorithms	Sequential	Random
Naïve Bayes	75,0 ± 8,3	64,2 ± 10,2
C4.5	75,8 ± 7,3	55,4 ± 7,8*
SVM	73,3 ± 4,9*	72,5 ± 6,8
1-NN	60,4 ± 9,8*	58,7 ± 8,8*
3-NN	64,2 ± 9,1*	59,6 ± 7,3*
5-NN	67,1 ± 7,9*	57,9 ± 8,2*
7-NN	68,3 ± 8,8*	55,4 ± 11,8*

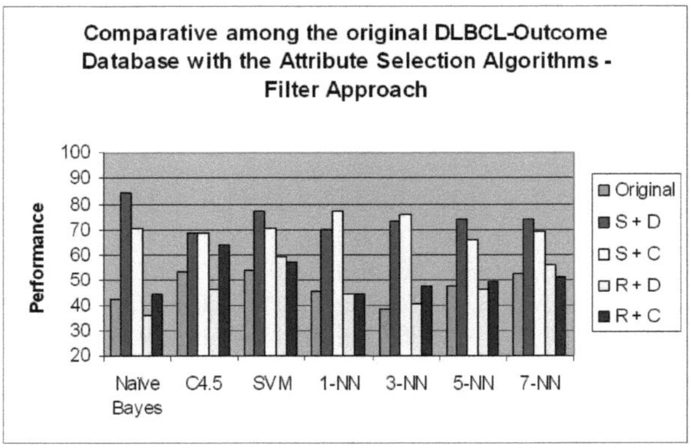

Fig. 1.8. Comparative among the original DLBCL-Outcome database with the attribute selection algorithms belonging result the filter approach

the 7-NN algorithm presented worse results in the sequential and random search method but just in the dependency evaluation measure.

Wrapper Approach

Using the wrapper approach the results were obtained on the subsets according to Tables 1.6 and 1.7. The first one shows the results obtained by the sequential search (S) method and the later presents the results using random search (R) method in each one of the databases. In Table 1.7 the obtained results of the database DLBCL-Outcome are presented. In the (S) search method the 1-NN algorithm had worse statistics results compared with the base algorithm. But, in the Random search method the 3-NN algorithm stood out with the best result statistic evaluation.

One of the objectives of the attribute selection is to improve the behavior of the algorithms, mainly in the cases where the rate of classification obtained on the database with all the attributes was low. Its application improved the success of the classifier in the case where all attribute subset were used.

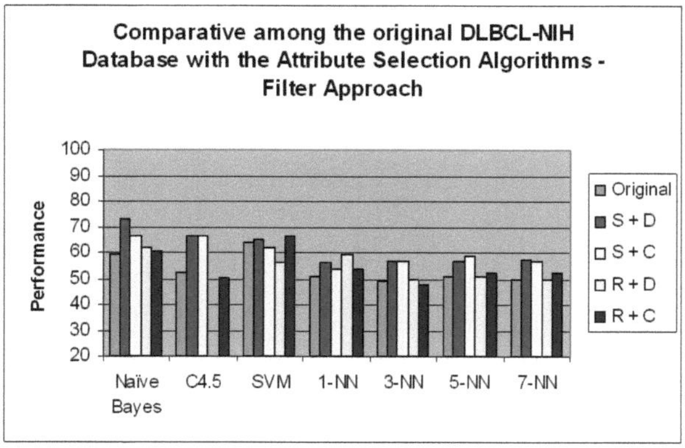

Fig. 1.9. Comparative among the original DLBCL-NIH database with the attribute selection algorithms belonging result the filter approach

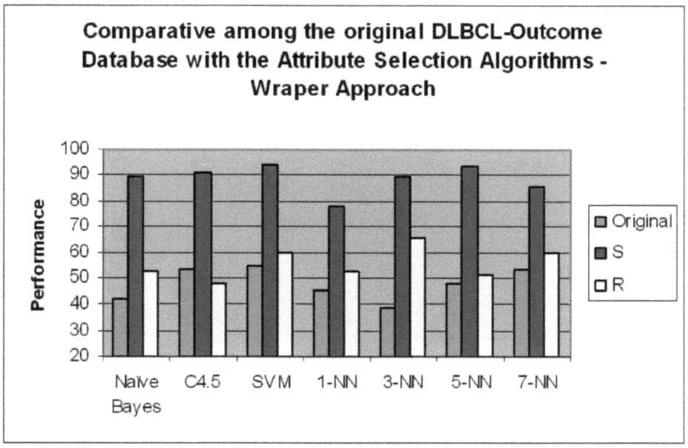

Fig. 1.10. Comparative among the original DLBCL-Outcome database result with the attribute selection algorithms belonging the wrapper approach

The figures 1.8 and 1.9 shows a comparative one among the classification algorithms results when used in them database with all of the attributes and when applied in the attributes subsets generated by the attribute selection algorithms using the filter approach.

The figure 1.10 and 1.11 shows a comparative one among the classification algorithms results when used in them database with all of the attributes and when applied in the attributes subsets generated by the attribute selection algorithms using the wrapper approach. Your performance is clearly superior in the best cases, independent of the classification algorithm that is being used.

Besides, the figures 1.12 and 1.13 shows a comparison of the classifiers performance when used in the original database and when applied in the best and worse cases.

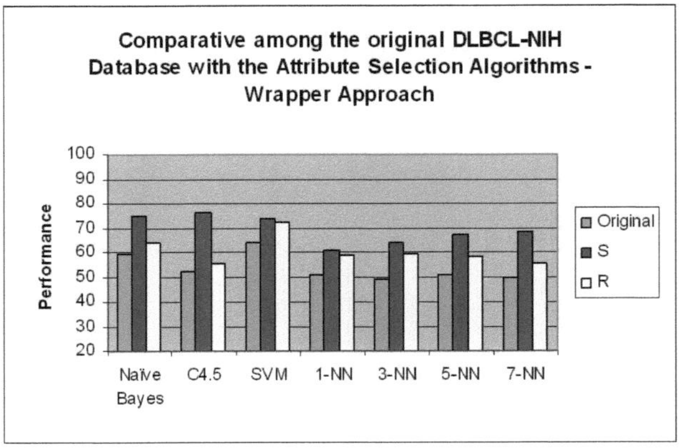

Fig. 1.11. Comparative among the original DLBCL-NIH database result with the attribute selection algorithms belonging the wrapper approach

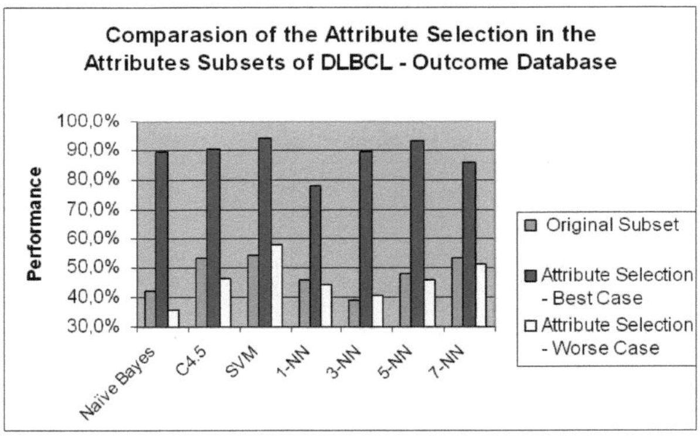

Fig. 1.12. Comparison of the attribute Selection Methods in the attribute Subsets of the DLBCL-Outcome database in the Best Case and in the Worst Case

Comparing the results of database classification when using all the attributes and when using only those attribute subsets generated by the attribute selection algorithms, one can observe that the success rate is clearly superior in the later cases, independent of the classification algorithm that is being used. In the worst cases, the results are very similar to the situation where all of the attributes are used. Therefore, we conclude that the use of the attribute selection is an advisable preprocessing step for bioinformatics data.

A more detailed analysis of the attributes selected (genes) reveals that some of them were selected in most of the cases. This fact can be used to look for their relationship with the disease. After doing an analysis of each experiment, the average of the classification rate of the algorithms applied to both databases was made.

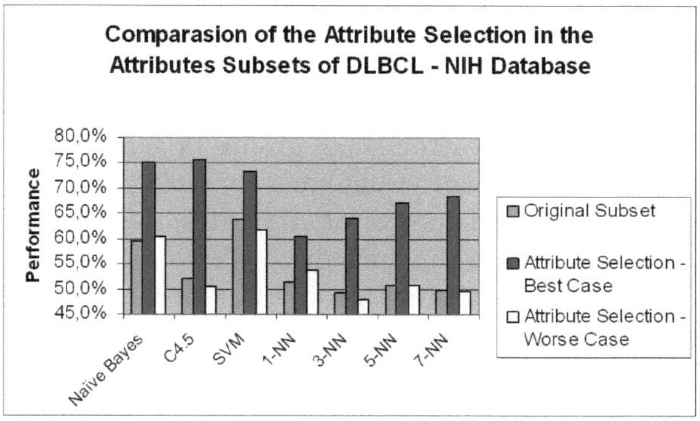

Fig. 1.13. Comparison of the attribute Selection Methods in the attribute Subsets of the DLBCL-NIH database in the Best Case and in the Worst Case

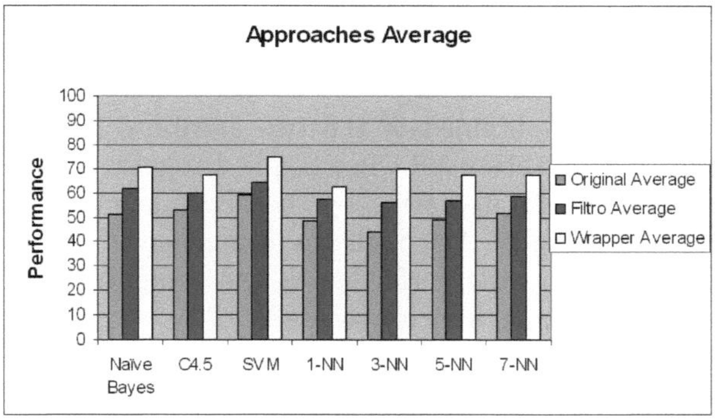

Fig. 1.14. Average of the results of the original database versus the application of the approaches for attribute selection

Figure 1.14 show the results obtained through the arithmetic average of the two databases. It can be noticed that the obtained average of the classification algorithms results in the databases with all the attributes was inferior to the result obtained by the attribute selection algorithms. The wrapper approach presents the best results in all the classification algorithms.

1.5 Conclusion

Machine Learning techniques are, indeed, used more often to help solving problems in Molecular Biology, due to their capacity of producing useful hypothesis through automatic learning from available data. Classification is one of the more common tasks in Data Mining since it allows the prediction of the class of a new example, based on its attributes.

Data from DNA micro-arrays are composed of a large number of genes (attributes) and a small number of samples. These characteristics of gene expression data makes hard for data mining algorithms to achieve good performance. Besides that, it is in known in Molecular Biology that some characteristics (like cancer arising) are to appear when a small groups of genes become associated.

In Data Mining, there are many techniques for data preprocessing as the attribute selection. It reduces the number of attributes, removes irrelevant, redundant, or noisy data, and brings the immediate effects for applications: speeding up a data mining algorithm, improving mining performance such as predictive accuracy and result comprehensibility.

The experimenters results accomplished in microarray data, it has been observed that the use of attribute selection, despite the huge number of attributes and small number of samples, has lead to a better performance of the classifiers in almost all cases and schemes on both datasets.

The exclusion of attributes improved the quality results obtained. Even in the worst cases, the classifier performance was superior the one of when applied on the databases with all the attributes. Besides, there was a great reduction in the amount of attributes selected, mainly, when it was applied the evaluation measures with the sequential search.

Comparing the results obtained of the two attribute selection approaches, the wrapper approach was observed when applied together with the sequential search produced best results, following by the dependency evaluation measure belonging the filter approach. Although the difference of the classifiers result has been small in what says the performance, the computational cost was very superior when the wrapper approach was used.

In general, the execution of the algorithms belong the filter approach had a time of processing in the order of seconds to minutes. Already the algorithms belonging the wrapper approach had a time of processing of the order of hours to days of processing, what can, in some cases, he becomes unviable your application.

It is also important to highlight that the database in study are bases that study the patient's survival, after a treatment period. However, just the data related to the patient's genes form the database. In that way, the other characteristics, that are related in the patient's survival and that can have great influences in your final result, they are not part of the database.

References

1. Alizadeh, A., et al.: Distinct types of diffuse large B-cell Lymphoma Identified by gene expression profiling. Nature 4051, 503–511 (2000)
2. Baldi, P., Long, A.D.: A Bayesian Framework for the Analysis of Microarray Expression Data: Regularized t-Test and Statistical Inferences of Gene Changes. Bioinformatics 17(6), 509–519 (2001)

3. Bicciato, S., Pandin, M., Didone, G., di Belle, C.: Analysis of an Associative Memory Neural Network for Pattern Identification in Gene Expression Data. In: BIOKDD 2001: Workshop on Data Mining in Bioinformatics (with SIGKDD 2001 Conference), pp. 22–30 (2001)
4. Bolshakova, N., Azuaje, F., Cunningham, P.: An integrated tool for microarray data clustering and cluster validity assessment. Bioinformatics 21(4), 451–455 (2005)
5. Borges, H.B., Nievola, J.C.: Attribute Selection Methods Comparison for Classification of Diffuse Large B-Cell Lymphoma. In: The 4th International Conference on Machine Learning and Applications - ICMLA 2005, Los Angeles, vol. 1, pp. 201–206 (2005)
6. Borges, H.B.: Redução de Dimensionalidade em Bases de Dados de Expressão Gênica. In: Dissertação (Mestrado em Informãtica Aplicada). p. 123, PPGIA - Pontificia Universidade Católica do Paran'a - PUCPR (2006)
7. Borges, H.B., Nievola, J.C.: Gene-finding as an Attribute Selection Task. In: Proceedings of the 6th IEEE International Conference on Computer and Information Science, vol. 1, pp. 537–542. IEEE Press, Los Alamitos (2007)
8. Brank, J., Grobelnik, M., Frayling, N., Mladenic, D.: Interaction of feature selection methods and linear classification models. In: Proceedings of 19th Conference on Machine Learning, Workshop on Text Learning (2002)
9. Brown, M.P.S., Grundy, W.N., Lin, D., Cristianini, N., Sugnet, C., Ares, Jr. M., Haussler, D.: Support Vector Machine Classification of Microarray Gene Expression Data, Technical Report UCSC-CRL-99-09, Department of Computer Science, University of California, Santa Cruz, Santa Cruz
10. Brown, M.P.S., Grundy, W.N., Lin, D., Cristianini, N., Sugnet, C.W., Furrey, T.S., Ares, Jr. M., Haussler, D.: Knowledge-Based Analysis of Microarray Gene Expression Data by using Support Vector Machin. In: PNAS, vol. 97(1), pp. 262–267 (January 4, 2000)
11. Butte, A.: The Use and Analysis of Microarray Data. Nature Reviews | Drug Discovery 1, 951–960 (2002)
12. Caropreso, M.F., Matwin, S., Sebastiani, F.: A learner-independent evaluation of the usefulness of statistical phrases for automated text categorization. In: Chin, A.G. (ed.) Text Databases and Document Management: Theory and Practice, pp. 78–102. Idea Group Publishing, Hershey (2001)
13. Debole, F., Sebastiani, F.: Supervised term weighting for automated text categorization. In: Proc. of SAC 2003, 18th ACM Symposium on Applied Computing, Melbourne, US, pp. 784–788 (2003)
14. Fayyad, U.M., et al.: KDD for science data analysis: issues and examples. In: Second International Conference on Knowledge Discovery and Data Mining, Portland, Oregon. AAAI Press, Menlo Park (1996)
15. Golub, T., et al.: Molecular classification of cancer: class discovery and class prediction by gene expression monitoring. Science 286, 531–537 (1999)
16. Hall, M.A.: Correlation-based Feature Selection for Machine Learning, p. 198. Thesis. Waikato University (1999)
17. Hall, M.: Correlation-based feature selection for discrete and numeric class machine learning. In: Proceedings of the Seventeenth International Conference on Machine Learning, pp. 359–366 (2000)
18. Hanczar, B., Courtine, M., Benis, A., Hannegar, C., Clement, K., Zucker, J.-D.: Improving Classification of Microarray Data Using Prototype-Based Feature Selection. SIGKDD Explorations 5(2), 23–30

19. Huang, D., Chow, T.W.S., Ma, E.W.M., Li, J.: Efficient selection of discriminative genes from microarray gene expression data for cancer diagnosis. IEEE Transactions on Circuits and Systems 52(9), 1909–1918 (2005)
20. Hautaniemi, S., Yli-Harja, O., Astola, J., Kauraniemi, P., Kallioniemi, A., Wolf, M., Ruiz, J., Mousses, S., Kallioniemi, O.-P.: Analysis and Visualization of Gene Expression Microarray Data in Human Cancer Using Self-Organizing Maps. Machine Learning 52, 45–66 (2003)
21. Kerr, M.K., Martin, M., Churchill, G.A.: Analysis of Variance for Gene Expression Microarray Data. Journal of Computational Biology 7(6), 819–837 (2000)
22. Kohavi, R., John, G.H.: The Wrapper Approach. In: Liu, H., Motoda, H. (eds.) Feature Extraction, Construction and Selection: a data mining perspective, pp. 33–49 (1998)
23. Liu, H., Motoda, H.: Feature Selection for Knowledge Discovery and Data Mining. Kluwer academic Publishers, Dordrecht (1998)
24. Liu, H., Motoda, H., Yu, L.: The Handbook of Data Mining, pp. 409–423. Lawrence Erlbaum Associates, Mahwah (2003)
25. Liu, H., Setiono, R.: A Probabilistic Approach to Feature Selection: a Filter Solution. In: Proc. of the 13th Int. Conf.on Machine Learning, pp. 319–327. Morgan Kaufmann, San Francisco (1996)
26. Liu, H., Yu, L.: Toward Integrating Feature Selection Algorithms for Classification and Clustering. IEEE Transactions on Knowledge and Data Engineering 17(4), 491–502 (2005)
27. Long, P.M., Veja, V.B.: Boosting and Microarray Data. Machine Learning 52, 31–44 (2003)
28. Miller, L.D., Long, P.M., Wong, L., Mukherjee, S., McShane, L.M., Liu, E.T.: Optimal Gene Expression Analysis by Microarrays. Cancer Cell 2, 353–361 (2002)
29. Molina, L.C., Belanche, L., Nebot, A.: Feature Selection Algorithms: A Survey and experimental Evaluation. Technical Report LSI-02-62-R, Universidade Politécnica de Catalunya, Barcelona, Espanha (2002)
30. Pagano, M., Gauvreau, K.: Princípios de Bioestatísica. São Paulo: Pioneira Thonson Learning (2004)
31. Piatetsky-Shapiro, G., Khabaza, T., Ramaswamy, S.: Capturing Best Practice for Microarray Gene Expression Data Analysis. In: Proceedings of the Ninth ACM SIGKDD International Conference on Knowledge Discovery and Data Mining, pp. 407–415 (2003)
32. Quackenbush, J.: Computational Analysis of Microarray Data. Nature Reviews | Genetics 2, 418–427 (2001)
33. Quinlan, J.R.: C4.5: Programs for Machine Learning. Morgan Kaufmann Publishers, San Mateo (1993)
34. Rosenwald, A., et al.: The use of molecular profiling to predict survival after chemotherapy for diffuse large-B-cell lymphoma. N. Engl. J. Med. 346(25), 1937–1947 (2002)
35. Rubinstein, B.I.P., McAuliffe, J., Cawley, S., Palaniswami, M., Ramamohanarao, K., Speed, T.P.: Machine Learning in Lowlevel Microarray Analysis. ACM SIGKDD Explorations Newsletter 5(2), 130–139 (2003)
36. Salton, G., McGill, M.J.: Introduction to Modern Retrieval. McGraw-Hill Book Company, New York (1983)
37. Shannon, W., Culverhouse, R., Duncan, J.: Analyzing microarray data using cluster analysis. Pharmacogenomics 4(1), 41–51 (2003)
38. Sheng, Q., Moreau, Y., De Bart, M.: Biclustering Microarray Data by Gibbs Sampling. Bioinformatics 9(2), 196–205 (2003)

39. Shipp, M., et al.: Diffuse large B-cell lymphoma outcome prediction by gene expression profiling and supervised machine learning. Nature Medicine 8(1), 68–74 (2002)
40. Soucy, P., Mineau, G.W.: Beyond TFIDF weighting for text categorization in the vector space model. In: International Joint Conference on Artificial Intelligence, Edinburgh, Scotland, UK, pp. 1130–1135 (2005)
41. Tibshirani, R., Hastie, T., Eisen, M., Ross, D., Botstein, D., Brown, P.: Clustering Methods for the Analysis of DNA Microarray Data, Technical report, Department of Health Research and Policy, Stanford University (1999)
42. Xing, E.P., Jordan, M.I., Karp, R.M.: Feature Selection for High-Dimensional Genomic Microarray Data. In: Proceedings of the Eighteenth International Conference on Machine Learning, pp. 601–608 (2001)
43. Yang, J., Honavar, V.: Feature subset selection using a genetic algorithm, Iowa State University Technical Report TR 97-02a.YA (1997)
44. Yang, Y., Pedersen, J.O.: A comparative study in feature selection on text categorization. In: Fishe, D.H. (ed.) Proceedings of ICMLA 1997, 1st International Conference on Machine Learning, Nashville, US, pp. 412–420 (1997)
45. Yu, L., Liu, H.: Redundancy Based Feature Selection for Microarray Data. In: Proceedings of the Tenth ACM SIGKDD Conference on Knowledge Discovery and Data Mining, pp. 737–742 (2004)
46. Witten, I.H., Ian, H., Frank, E.: Data Mining: Practical machine learning tools and techniques, 2nd edn. Morgan Kaufmann, San Francisco (2005)

2

Preprocessing Techniques for Online Handwriting Recognition

Bing Quan Huang, Y.B. Zhang, and M.-T. Kechadi

School of Computer Science & Informatics, University College Dublin, Belfield, Dublin 4, Ireland
bingquan.huang@ucd.ie

As a general first step in a recognition system, preprocessing plays a very important role and can directly affect the recognition performance. This Chapter proposes a new preprocessing technique for online handwriting. The approach is to first remove the hooks of the strokes by using changed-angle threshold with length threshold, then filter the noise by using a smoothing technique, which is the combination of the Cubic Spline and the equal-interpolation methods. Finally, the handwriting is normalised. Section 2.1 introduces the problems and the related techniques of the preprocessing for online handwritten data. Section 2.2 describes our preprocessing approach for online handwritten data. The experimental results with discussions are showed in Section 2.3. The summary of this chapter is given in the last section.

2.1 Introduction

Most digital tablets have low-pass hardware filters. Thus, when such devices are used to capture handwriting strokes, the shapes of the strokes present jagged forms. On the other hand, most handwriting strokes contain some hooks and duplicated sampled points caused by hesitant writing. Furthermore, some points of a stroke may be missing, and wild points might exist in a stroke, etc. Generally, such noise information influences the exploration of the profile of the handwritten strokes in such a way as to influence further processes, such as feature extraction and classification. In contract to most of time series applications, handwriting is much more complex. For example, in the preprocessing stage of energy load forecasting [8, 13], one recurrent neural network (RNN) is applied to smooth the data that contains a small amount of noise. However, the RNN cannot perform regression in the term of interpolating the missing data. In order to improve recognition rates, it is necessary to remove the noisy data, interpolate the missing points and normalise strokes' sizes (symbols, words and expressions).

Many techniques have been applied to pre-process online handwritten data [4, 10, 15, 16]. These experimental results show that the recognition rate can be

N. Nedjah et al. (Eds.): Intelligent Text Categorization and Clustering, SCI 164, pp. 25–45.
springerlink.com

improved by at least 6% when using appropriate preprocessing techniques. However, most of them may not efficiently deal with the variety of the writing styles (such as different writing speeds from different writers). Deepu et al. [2] used Gaussian technique to filter handwritten data. However, the Gaussian cannot interpolate the points [1]. Ramsay [12] used Spline to filter handwritten data when giving sampled points. Although splines [3] can interpolate the missing points, they still cannot efficiently remove some of the variety of writing styles. Similarly, there are difficulties in using Wavelet to deal with some strokes of variable lengths [16]. Chew et al. [9] converted the handwriting strokes into a image, and then used wavelet to filter the image. Therefore, they actually used the wavelet technique to smooth images (not to filter online data).

In order to avoid distortion of the handwriting and solve the above limitations, we propose a new preprocessing technique in this study. The approach firstly removes the hooks in the strokes, then uses the Cubic Spline with equal-interpolation methods to perform the filtering in order to reduce the noise. Finally, the online handwritten symbol is normalised. The rest of this chapter gives the details of the this preprocessing approach.

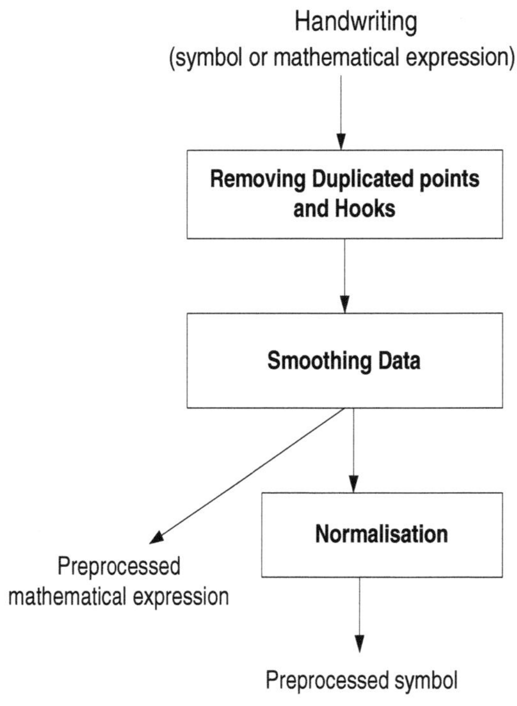

Fig. 2.1. Preprocessing steps

2.2 Preprocessing Approach

The goal of preprocessing is to discard irrelevant information that can negatively affect the recognition. Generally, the irrelevant information includes duplicated points, wild points, hooks, noise, etc. Our approach is summarised in Figure 2.1. Given an online handwritten symbol or expression, the approach removes the duplicated points and hooks, smoothes the data, and performs normalisation. As a result, the profile of online handwritten symbols can be easily and accurately detected. In the following we describe each of these steps.

2.2.1 Elimination of Duplicated Points and Hooks

Removing Duplicated Points

Some authors [4] use a filtering approach to insert some space between the duplicated points in the arc. Here, the duplicated points are removed by checking whether the coordinates of any two points are the same. If they are in the same position, one of them is kept and others are removed.

Removing Hooks

Usually, hooks occur at the beginning and/or at the end of a stroke (see Figures 2.2 and 2.7). In these figures, some jagged points are marked by red circles. They are due to inaccuracies in rapid pen-down/up detection and erratic hand-motion. In addition, hooks are usually characterised by their small length and their large changed-angular variations [4]. Based on this information (their locations, their small length and sharp turn angles), an approach to remove the hooks is detailed. The algorithm firstly uses the equal-length technique to interpolate and adjust points, then uses the detection method of the sharp points to remove the hooks. Let a stroke S with N points be:

$$S = \{p_0(x_0, y_0), \cdots, p_{N-1}(x_{N-1}, y_{N-1})\} \tag{2.1}$$

The three steps for the elimination of the hooks are described in the following.

- **Interpolating points:**
  When many points are missing between two consecutive points p_i and p_{i+1}, it is necessary to regenerate and add some points between them. Given a length d, the process iteratively inserts a new point every a length d along the curve that is from point p_i to point p_{i+1}. d is a constant during interpolating and adjusting points. d is not usually bigger than the average length which is computed by Eq.(2.2). The selection of the length d relies on the input device. In our experiment, d is half of the average length of a stroke for the UNIPEN dataset. A new point $p_{new}(x', y')$ is inserted according to the Eq.(2.4).

$$L_e = \frac{\sum_{i=1}^{N-1} len(i-1, i)}{N} \tag{2.2}$$

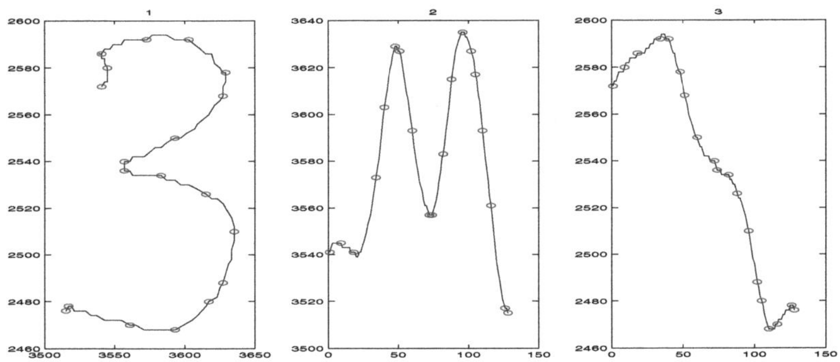

Fig. 2.2. The raw handwriting data presented by wavelets. (1) shows digit 3. (2) shows x-wavelet, and (3) shows y-wavelet of the digit.

where $len(a, b)$ is a function. It is used to calculate the distance between the two points $p_a(x_a, y_a)$ and $p_b(x_b, y_b)$.

$$len(a, b) = \sqrt{(x_a - x_b)^2 + (y_a - y_b)^2} \tag{2.3}$$

$$x_{new} = \begin{cases} x_i + \sqrt{\frac{d^2}{k^2+1}} & if \;\; x_i < x_{i+1} \\ x_i - \sqrt{\frac{d^2}{k^2+1}} & if \;\; x_i > x_{i+1} \\ x_i & if \;\; x_i = x_{i+1} \end{cases}$$

$$y_{new} = \begin{cases} kx + y_i - kx_i & if \; x_i \neq x_{i+1} \\ y_i + d & if \;\; y_i < y_{i+1} \; \& \; x_i \neq x_{i+1} \\ y_i - d & if \;\; y_i > y_{i+1} \; \& \; x_i \neq x_{i+1} \end{cases} \tag{2.4}$$

and where k is the slope of the line defined by the two points p_i and p_{i+1}, defined as follows:

$$k = \frac{y_{i+1} - y_i}{x_{i+1} - x_i} \quad if \;\; x_{i+1} \neq x_i \tag{2.5}$$

If the given d is greater than the length of the curve between the two points p_i and p_{i+1}, the process first searches for the point p_{i+n} from which the curve to the point p_i is longer than length d. Once the point p_{i+n} is found, a new point is added into the curve between two points p_i and p_{i+n} by using Eq.(2.4). Then the old points on the curve from p_i to the new added point are removed. Considering the new point added being a new start point, the process continues to interpolate or adjust points based on length d. When the last point of the stroke is reached, the last one is kept and the process for adding and adjusting points stops.

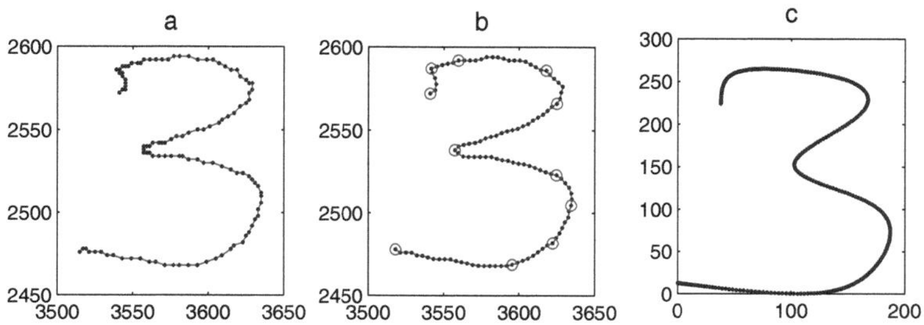

Fig. 2.3. Preprocessing. (a) raw data, (b) the data with the control points after removing the hooks and duplicated points, and (c) the normalised data. The control points are marked by the red circles. The points are represented by blue dots.

Figure 2.3 shows an example of digit 3 preprocessed by equal length method, when d is equal to the average length of the curve, which is obtained by Eq.(2.2. Figure 2.3(b) shows that the amount of redundant information (e.g., the duplicated points, wild points) were removed, and it also shows that the curve lengths between any two consecutive points are equal.

- **Detection of sharp points:**
As mentioned above, hooks are at very sharp points and often occur when the large changed-angle of the writing is formed. Figures 2.2 and 2.7 show the locations of the hooks which are always at the sharp points, and the length of the hooks, which are often shorter. Therefore, it is very important to find the sharp points, and to remove the hooks from a stroke. A sharp point is at the peak of the wavelet where the writing direction has changed. The algorithm for exploring sharp points is developed based on the changed-angles of pen motion. Firstly, according to the given stroke S, the algorithm calculates the slope angles of the lines that are defined by any two consecutive points. Secondly, the algorithm calculates the changed angles based on these slopes. Note that a changed-angle is the angle between two consecutive lines.
In the values of these changed-angles, a turn (or joint) values between the subset of the consecutive increasing changed-angle values and the subset of the consecutive decreasing changed-angle values can be identified, which represents a sharp point. In addition, the beginning point and end point of a stroke are sharp points. Therefore, the sharp points in a stroke can be calculated by the turn (or joint) values with the start and end points. Consider the sharp points are stored in V. Given a stroke, Algorithm 1 describes how to search for the sharp points.
In Algorithm 1, $\alpha_{i,i+1}$ is the angle between two lines: one is defined by the points (x_i, y_i) and (x_{i+1}, y_{i+1}), the other is defined by the points (x_{i+1}, y_{i+1}) and (x_{i+2}, y_{i+2}), and Δ is a variable which determines whether an angle is a turn angle. If Δ is not positive, then the writing direction is altered. Figure 2.4 shows an example of the detection of sharp point p_{i+2} in the stroke.

Algorithm 1. Sharp Point Detection

1. add the first point p_1 to V.
2. Calculate the slope angles of the lines, each of which is defined by two consecutive points in strokes. These angles can be written as $\alpha = \{\alpha_{1,i}, \cdots, \alpha_{N-1,N}\}$.
3. %where i is the index of the points.
4. $i = 1$
5. **while** $i < N$ **do**
6. $\quad \theta_{i,i+1} = \alpha_{i,i+1} - \alpha_{i+1,i+2}$,
7. $\quad$ **if** $\theta_{i,i+1} = 0$ **then**
8. $\qquad i{+}{+}$
9. $\qquad$ Go to step 5
10. $\quad$ **end if**
11. $\quad \Delta = \theta_{i,i+1} * \theta_{i-1,i}$
12. $\quad$ **if** $(\Delta <= 0) \& (\theta_{i-1,i} \neq 0))$ **then**
13. $\qquad$ The pen moving direction is changed at point p_{i+1}. Therefore, add the first point p_{i+1} to V.
14. $\quad$ **end if**
15. $\quad i{+}{+}$
16. **end while**
17. add the last point p_N to V.

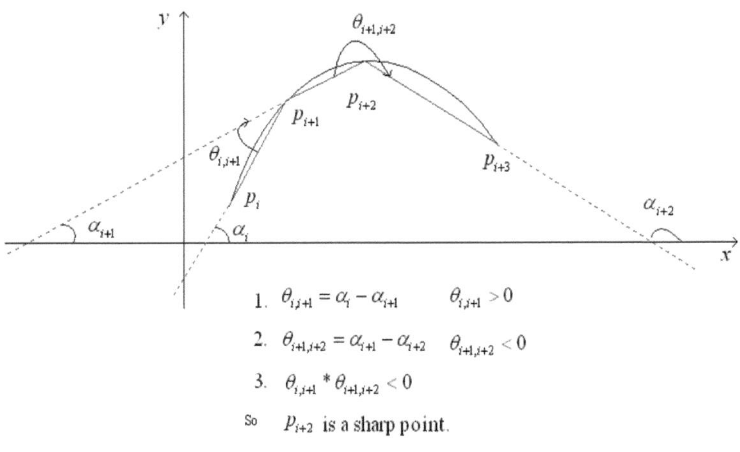

Fig. 2.4. Sharp point detection

- **Removing hooks:**
  After the sharp points have been detected, if there are more than two sharp points, the hooks of the stroke can be found based on two segments: one segment seg_b is between the first two sharp points, the other segment seg_e is between the last two sharp points. Let their slope angles be β_b and β_e respectively. Also let the length of segment seg_b or seg_e be l_{seg} respectively, which can be computed by the Eq.(2.3).

  In order to check whether these two segments are hooks, two segment have to be defined: one segment seg_{b+1} is between the first second and third sharp points, the other segment seg_{e-1} is between the second and third last sharp points. The angle λ between segments seg_b and seg_{b+1} or between segments seg_e and seg_{e-1} can be obtained by

$$\lambda = \begin{cases} |\beta_b - \beta_{b+1}| \\ |\beta_e - \beta_{e-1}| \end{cases}$$

Thus, if the length l_{seg} and the angle λ of segment seg_b or seg_e match the following conditions, the segment is a hook.

$$\lambda \leq threshold_{angle} \quad and \quad l_{seg} \leq threshold_{len}$$

Here, $threshold_{angle}$ is set to 90°, and $threshold_{len}$ is set to 3% of the diagonal line. The diagonal line of a stroke S is computed based on the bounding box of the stroke. The bounding box is obtained by the following formula:

$$\begin{aligned} Min_x &= \min_{0 \leq i \leq N-1} x_i \\ Min_y &= \min_{0 \leq i \leq N-1} y_i \\ Max_x &= \max_{0 \leq i \leq N-1} x_i \\ Max_y &= \max_{0 \leq i \leq N-1} y_i \end{aligned} \tag{2.6}$$

The length of the diagonal line for this stroke is computed by the following formula:

$$L_d = \sqrt{w^2 + h^2} \tag{2.7}$$

where

$$\begin{aligned} w &= Max_x - Min_x \\ h &= Max_y - Min_y \end{aligned} \tag{2.8}$$

Figure 2.3b shows that the hook at the end of the stroke was efficiently removed, compared with Figure 2.3a.

2.2.2 Smoothing Data

Although the equal-length sampling method can smooth the data, the result is not highly accurate. Therefore, the handwriting data needs to be filtered again. Here the equal-length technique and the Cubic Spline are used together to filter the handwriting data. The process of our method for smoothing data employs the Cubic Spline and, then applies the equal-length method to filter handwriting data as shown in Figure 2.5.

Consider a set of data points $\{p_k(x_k, y_k), k = 0, 1, \cdots, N-1\}$ with $a = x_0 <, \cdots, < x_N$ is given to the function $F(x)$. The function $F(x)$ is a Cubic Spline, when there exists N cubic polynomials $F_k(x)$ with coefficients $f_{k,i}, 0 \leq i \leq 3$ which are represented by E.q (2.9).

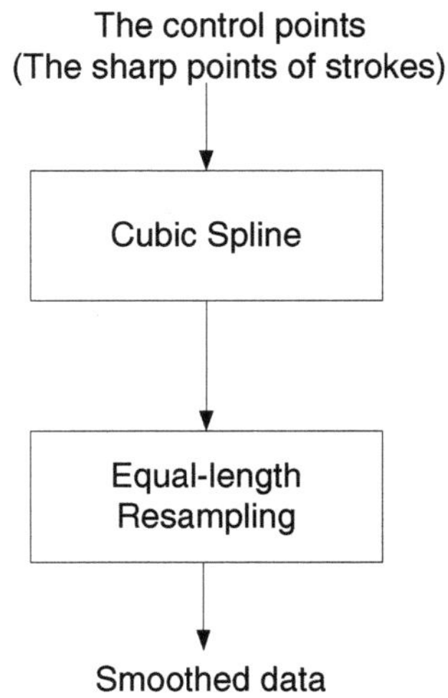

Fig. 2.5. Smoothing Process

$$F(x) = F_k(x) = \sum_{i=0}^{3} f_{k,i}(x - x_k)^i \forall x \in [x_k, x_{k+1}] \quad 0 \le k \le N-2 \qquad (2.9)$$

$$F_k(x) = y_k \quad 0 \le k \le N-1 \qquad (2.10)$$

$$F_k(x) = F_{k+1}(x_{k+1}) \quad 0 \le k \le N-3 \qquad (2.11)$$

$$F_k^{'}(x_{k+1}) = F'_{k+1}(x_{k+1}) \quad 0 \le k \le N-3 \qquad (2.12)$$

$$F_k^{''}(x_{k+1}) = F''_{k+1}(x_{k+1}) \quad 0 \le k \le N-3 \qquad (2.13)$$

The detail of the Cubic Spline can be found in [3].

A Cubic Spline is a spline constructed of piecewise third-order polynomials which pass through a number of the control points. Given a number of control points of a stroke, the Cubic Spline can interpolate a number of the points between any two consecutive points. The sharp points of a stroke are regarded as the control points, which can be obtained by the method introduced by Algorithm 1. Once the stroke has been smoothed by using the Cubic Spline, the equal-length method which is mentioned in Section 2.2.1, is applied to re-sample the stroke. Figure 2.3*c* shows the digit 3 that is smoothed by this smoothing technique. Figure 2.6 shows the deviated angles after the deduction step of duplicated points and hooks, and the smoothing step. Figure 2.6 (2) especially shows that the handwriting data smoothed by this method is much smaller than the other two. The above techniques, which are used for removing duplicated points and hooks, and for filtering handwriting data, are all based on a stroke. Therefore, the irrelevant information of a symbol or an expression can be discarded by using the same process with the same technique to filter the irrelevant information of each stroke.

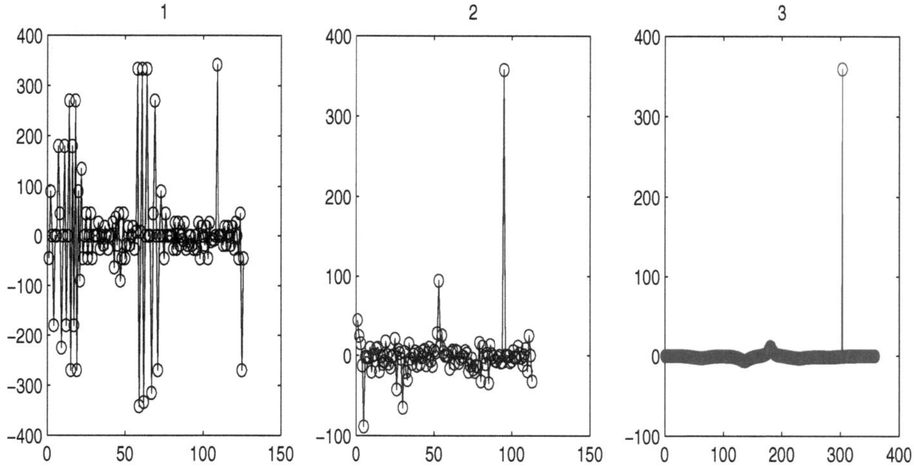

Fig. 2.6. The changed-angles between two lines from digit 3 which are shown in Figure 2.3. (1) is related to Figure 2.3*a*, (2) is related to Figure 2.3*b*, and (3) is relative to Figure 2.3*c*.

2.2.3 Normalisation

The goal here is to remove some of the variations of handwriting styles and to simplify the shapes of symbols. Usually, it includes two of the important procedures: scaling and transition. The aim of scaling and transition is to scale all the symbols into the bounding box with the same location and size. Thus, the feature of the symbols can be easily discovered.

Suppose a symbol consists of a sequence of points. It is presented as

$$S^{'} = p_i(x_i, y_i), i = 1, 2, \cdots, N \tag{2.14}$$

In order to scale symbol $S^{'}$ into the size where the width and the height are W' and H' respectively, the original width and height of the symbol have to be calculated. They can be computed by Eq. (2.8). Thus, a system S' can be scaled by the following equation:

$$\begin{aligned} x_i &= x_i * \frac{W'}{w} \quad i = 1, 2, \cdots, N, \\ y_i &= y_i * \frac{H'}{h} \quad i = 1, 2, \cdots, N, \end{aligned} \tag{2.15}$$

However, when the symbol shrinks ($W' < w$ and $H' < h$), a new symbol might include some duplicated points. Therefore, if the symbol is shrunk and contains some duplicated points, the duplicated points have to be removed before translating the symbol.

Suppose a symbol $S' = \{p_0(x_0, y_0), \cdots, p_{N-1}(x_{N-1}, y_{N-1})\}$ will be translated into x_l, y_l. The transition (lx, ly) can be computed by

$$\begin{aligned} lx &= Min_x - x_l \\ ly &= Min_y - y_l \end{aligned} \tag{2.16}$$

where Min_x and Min_y can be obtained by Eq.(2.7). Thus, the symbol S' can be translated by the following equation:

$$\begin{aligned} x_i &= x_i - lx \quad i = 0, 1, \cdots, N-1 \\ y_i &= y_i - ly \quad i = 0, 1, \cdots, N-1 \end{aligned} \tag{2.17}$$

In this thesis, the average size (189 × 267) of all training samples is used to scale symbols, and then the symbols are shifted at the bottom-left of the pen-pad by setting Min_x and Min_y to $(0, 0)$, lx and ly respectively.

In Figure 2.3 the symbol is pre-processed. Figure 2.3(a) shows the raw data. Figure 2.3(b) shows the clear handwriting data and the control points, which are identified by the red circles. These control points are used by the Cubic Spline to regress the strokes. Figure 2.3(c) shows the normalised symbol whose size is

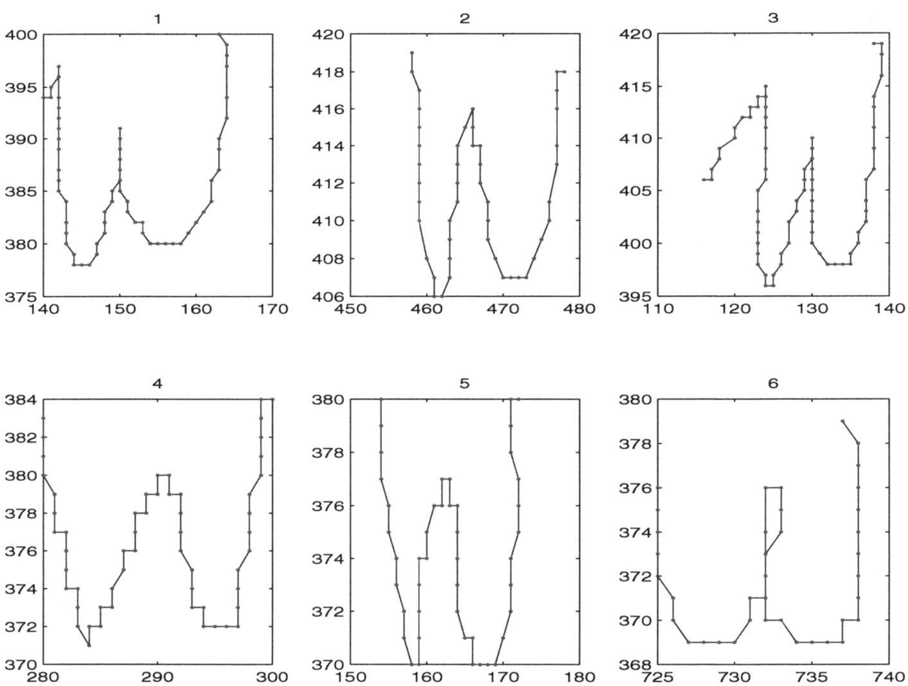

Fig. 2.7. The raw character w

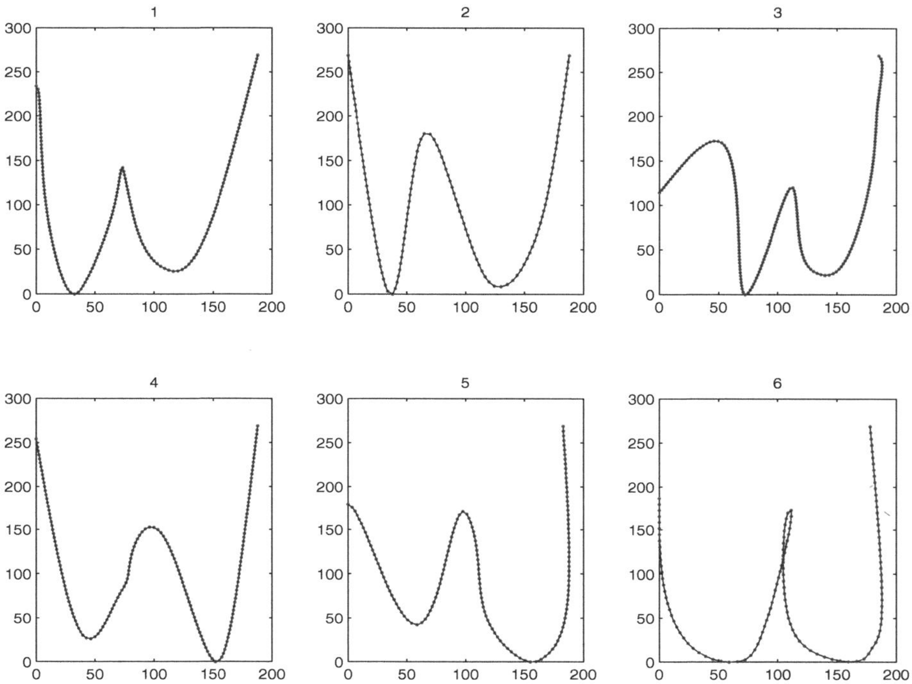

Fig. 2.8. The preprocessed character w

189 × 267, and also shows that the symbol is shifted to the left-bottom corner at (0,0). In addition, more examples are provided by Figure 2.7 and Figure 2.8. Figure 2.7 shows the original raw online handwritten symbol w. Figure 2.7 shows the online handwritten symbol w, after all of the procedures of the preprocessing has gone through the symbols. Also, the duplicated points and the hooks are eliminated, the symbol w is smoothed and normalised. In each of these two figures, the 5 samples of symbol w are respectively given index numbers 1, 2, 3, 4 and 5. The samples 1, 2, 3, 4 and 5 in Figure 2.7 are respectively preprocessed to respectively form the samples 1, 2, 3, 4 and 5 in Figure 2.8. In these two figures, the sampled points are represented by the red dots, and the blue lines are the links along the points of the stroke. In contrast to the hooks in the samples 1, 2 and 4 in Figure 2.8, the hooks in samples 1, 2, and 4 in Figure 2.7 are efficiently removed. Figure 2.7 shows the different sampling rates of the devices which generated different jagged forms of the symbol. However, Figure 2.8 shows that the symbols are efficiently reconstructed, and much irrelevant information is discarded. Figure 2.8 also shows that all of the symbols are located in are the same place and are the same size, because the scales of the axis (x and y) of all examples are the same.

2.3 Experiments and Discussion

Usually, the evaluation of the processing methods is carried out by a classifier or visualisation method (which was demonstrated in above sections). The former employes a machine to classify symbols, and the later is one of human-subjective evaluation methods. Therefore, the former is an objective method and preferred for evaluating preprocessing methods. Three sets of the experiments were independently carried out to recognise online handwritten symbols: digits, upper- and lower-case characters. The training and testing datasets of these types of symbols are from the UNIPEN database [5]. 11974, 20786 and 42492 samples of digits, uppercase and lowercase characters respectively were used as testing data. The training data includes 10000, 21840 and 37440 samples of digit, uppercase and lowercase characters, respectively.

In order to evaluate this preprocessing approach, for each set of the experiments, firstly the symbols were recognised without preprocessing. Then, the same symbols were recognition with preprocessing. The recognition process of the symbols without preprocessing is to firstly extract the features of the symbols, and then to classify the symbols. For the recognition of the symbols with preprocessing, the preprocessing for handwriting data was carried out before the feature extraction and classification. In the preprocessing stage, our approach was applied to remove the duplicated points and hooks, then filter the handwriting stroke, and finally normalise it. In this stage, the thresholds for removing hooks are 90°, 3% of changed angles and the diagonal line respectively (as mentioned above). The size for the normalisation is 189×267, where 189 and 267 are the average width and height respectively of all the training symbols.

The following features are used for the non-preprocessed and preprocessed symbols. The details of them and the extraction algorithm can be found in [7].

- Direction: indicating which eighth of a circle on the plane a stroke falls into.
- Re-curvature: characterising the ratio between the height of a stroke and the distance between its start and end points.
- Changed direction: being a measurement of the change in direction.
- Length: describing the length of a stroke.
- X-centre: giving the horizontal position of a stroke within the symbol. The closer it is to the right, the higher is the value.
- Y-centre: representing the vertical position of stroke within the symbol.
- Pen down/up: identifying when pen is down and lifted.
- Mass centre: a mass centre of a symbol
- Width/height ratio of symbols: The distortion may take place when a symbol is scanned during the normalisation step. To solve this problem, the ratio of the width and height of the original symbol is required.
- Zone information: The zones can be extracted by the following two steps: the map of a normalised symbol is firstly divided into a number of regions (blocks); each region is then addressed by x- and y-histogram features.

Two types of recognisers are independently used to classify symbols: one is based on Hidden Markov Models (HMM), the other is based on the hybrid HMM-MLP approach which combines Hidden Markov Models and a multilayer Perception artificial neural network. The details of them are provided in [6, 7].

In this experiment, the training data of each symbol was divided into 5 groups by using the K-mean algorithm to cluster their centres, and each group was constrained into about 150 samples. Then, a left-right HMM which consisted of 5 states and 5 multivariate Gaussian density functions was trained with each group of clustered samples over a period of 5 cycles by the Baum-Welch method [11]. Details of clustering data and of proposing and training the HMM classifier can be found in [7]. Tables 2.1, 2.2 and 2.3 respectively detail the recognition results of digits, uppercase and lowercase characters with and without the preprocessing step. Each table shows that a number of samples (patterns) are correctly/wrongly classified for each symbol. Each table shows the relative correctly recognition rates. The last row of each table summarizes the recognition results. Each table shows that the recognition rates of symbols with preprocessing are much higher than the rates of the same symbols without preprocessing. The tables also show that the recognition rates of some symbols without preprocessing are very low, such as digit 6 in Table 2.1, characters D,H, K and W in Table 2.2, characters h,m, u and w in Table 2.3. However, the recognition rates of these symbols with preprocessing are greatly improved (see the right-hand side of each table). Table 2.4 summaries the recognition rates of digits, uppercase and lowercase characters with preprocessing and without preprocessing from the pure HMM recognisers. The recognition rates were improved +40.59%, +64.28% and +56.01% for digits, uppercase and lowercase characters respectively.

For the hybrid HMM-MLP approach, the HMM-MLP classifier consists of one MLP and the same set of the HMMs for each type of symbol. The number of the left-right HMMs with training conditions and the data for each HMM of the hybrid classifier are as the same as the ones used in the HMM classifier (mentioned above). As mentioned in [7], the MLP of a hybrid recogniser accepts the new features that are combined with a number of HMM features and the

Table 2.1. Recognition rates with/without preprocessing of digits based on the HMM approach

		Without Preprocessing			With Preprocessing		
Digit	N.digits	N.correct	N.wrong	R.rate	N.correct	N.wrong	R.rate
0	1155	911	244	78.87%	1111	44	96.19%
1	676	370	306	54.73%	613	63	90.68%
2	1391	861	530	61.90%	1373	18	98.71%
3	1340	807	533	60.22%	1316	24	98.21%
4	1219	503	716	41.26%	1155	64	94.75%
5	1206	744	462	61.69%	1080	126	89.55%
6	1119	447	672	39.95%	1107	12	98.93%
7	1280	624	656	48.75%	1232	48	96.25%
8	1204	618	586	51.33%	1158	46	96.18%
9	1384	776	608	56.07%	1364	20	98.55%
Total	11974	6661	5313	55.63%	11509	465	96.12%

Table 2.2. Recognition rates with/without preprocessing of uppercase characters based on the HMM approach

sym-		Without Preprocessing			With Preprocessing		
bol	N.symb.	N.correct	N.wrong	R.rate	N.correct	N.wrong	R.rate
A	1170	348	822	29.72%	985	185	84.19%
B	764	255	509	33.37%	747	17	97.77%
C	856	459	397	53.62%	821	35	95.91%
D	817	106	711	12.99%	760	57	93.02%
E	1077	714	363	66.26%	913	164	84.77%
F	688	93	595	13.45%	608	80	88.37%
G	629	186	443	29.63%	573	56	91.10%
H	771	34	737	4.42%	694	77	90.01%
I	914	229	686	25.00%	790	124	86.43%
J	596	162	434	27.11%	550	46	92.28%
K	733	12	721	1.58%	636	97	86.77%
L	656	160	496	24.36%	622	34	94.82%
M	636	77	559	12.04%	606	30	95.28%
N	783	207	576	26.42%	665	118	84.93%
O	1421	197	1224	13.87%	1292	129	90.92%
P	781	207	574	26.56%	728	53	93.21%
Q	633	147	486	23.19%	600	33	94.79%
R	924	139	785	15.04%	871	53	94.26%
S	1000	480	520	47.99%	993	7	99.30%
T	1045	562	483	53.76%	856	189	81.91%
U	781	106	675	13.55%	708	73	90.65%
V	625	124	501	19.91%	556	69	88.96%
W	707	53	654	7.48%	666	41	94.20%
X	604	258	346	42.68%	535	69	88.58%
Y	636	75	561	11.74%	588	48	92.45%
Z	539	168	371	31.25%	533	6	98.89%
Total	20786	5555	15231	26.73%	18896	1890	90.91%

global features. A number pairs of HMM features correspond to the same number of the candidates. In paper [6], the authors confirm that 4 HMM features can achieve the highest recognition rates. However, the features extracted from the data without preprocessing for the HMMs might achieve low recognition rates. Thus, the actual label might not be represented by the four 4 HMM features. In addition, the objective of the experiment is not to find the best pairs of HMM which can achieve the highest recognition (as in [6]), but is to confirm the efficiency of the processing technique. Therefore, the beginning 8 HMM features provide more candidates than the 4 HMM features do. Therefore, the beginning 8 HMM features were selected to combine with global feature to form the new feature vector in this experiments for the HMM-MLP method. The MLP in each hybrid recogniser consists of three layers – input, hidden and output layers. 112, 144 and 144 input neurons respectively are in the 3 MLPs to recognise digits, uppercase and lowercase characters. Similarly, 10, 26 and 26 output neurons were

Table 2.3. Recognition rates with/without preprocessing of lowercase characters based on the HMM approach

sym-		Without Preprocessing			With Preprocessing		
bol	N.symb.	N.correct	N.wrong	R.rate	N.correct	N.wrong	R.rate
a	2770	1032	1738	37.24%	2480	290	89.53%
b	861	432	429	50.23%	777	84	90.24%
c	1349	159	1190	11.80%	1190	159	88.21%
d	1573	837	736	53.20%	1447	126	91.99%
e	4551	2186	2365	48.04%	4258	293	93.56%
f	1011	269	742	26.60%	805	206	79.62%
g	1128	229	899	20.27%	1019	109	90.34%
h	1725	96	1629	5.57%	1600	125	92.75%
i	2141	1075	1066	50.21%	2034	107	95.00%
j	625	186	439	29.76%	603	22	96.48%
k	958	173	785	18.04%	809	149	84.45%
l	2096	676	1420	32.27%	1513	583	72.19%
m	1256	93	1163	7.40%	1122	134	89.33%
n	2315	270	2045	11.68%	2089	226	90.24%
o	3454	1204	2250	34.87%	3144	310	91.02%
p	1358	231	1127	17.04%	1311	47	96.54%
q	845	358	487	42.38%	728	117	86.15%
r	1606	1128	478	70.24%	1358	248	84.56%
s	2112	1123	989	53.18%	2042	70	96.69%
t	2864	1520	1344	53.06%	2592	272	90.50%
u	1727	1	1726	0.06%	1274	453	73.77%
v	751	51	700	6.78%	614	137	81.76%
w	1043	4	1039	0.39%	974	69	93.38%
x	566	172	394	30.37%	464	102	81.98%
y	1068	409	659	38.27%	877	191	82.12%
z	739	102	637	13.87%	693	46	93.78%
Total	42492	14017	28475	32.99%	37817	4675	89.00%

Table 2.4. Recognition rates with/without preprocessing step based on the HMM classifiers

	Rec.Rate of Digits %	Rec.Rate of U.Chars.%	Rec.Rate of L.Chars.%
unpreproc.	55.63	26.73	32.99
preproc.	96.12	90.91	89.00
increased rate	+40.59	+64.28	+56.01

respectively in these 3 MLPs. 15, 20, 25, 30, 35, 40, 45, 50, 55 and 60 hidden neurons were used independently for each MLP. The Back-propagation (BP) [14] with a learning rate of 0.2, a maximum updating rate of 0.02, a tolerable MSE error of 0.1 and a maximum training cycle of 500, was used for each MLP.

Table 2.5. The increased recognition rates

	C.Rates of Non-preproc.Data %			C.Rates of prepro. Data %		
N. Hid.neuro.	Digits	U.Chars.	L.Chars.	Digits	U.Chars.	L.Chars.
15	85.32	75.15	71.13	92.23	90.07	84.07
20	85.62	75.97	72.54	95.96	91.3	89.31
25	85.95	76.56	73.09	96.59	91.41	90.11
30	87.36	78.26	73.15	97.33	91.63	91.32
35	86.8	78.4	75.31	96.44	91.74	91.51
40	85.77	77.74	75.16	96.36	91.03	90.91
45	83.76	77.09	74.18	96.26	90.96	90.88
50	83.48	76.87	72.13	96.06	88.54	88.27
55	82.71	75.1	72.71	96.05	86.03	85.88
60	77.25	73.67	72.26	95.9	85.98	85.83

Table 2.6. The recognition rates from the preprocessed and non-preprocessed data with different numbers of hidden neurons, based on the hybrid HMM-MLP approach

N. Hid.neuro.	digits %	U.Chars. %	L.Chars.%
15	6.91	14.92	12.94
20	10.34	15.33	16.77
25	10.64	14.85	17.02
30	9.97	13.37	18.17
35	9.64	13.34	16.2
40	10.59	13.29	15.75
45	12.5	13.87	16.7
50	12.58	11.67	16.14
55	13.34	10.93	13.17
60	18.65	12.31	13.57

After the HMM-MLPs were trained, they were used to classify the symbols with and without preprocessing. Table 2.6 shows the recognition rates of digits, uppercase and lowercase characters with preprocessing and without preprocessing from the HMM-MLP recognisers with a varied number of hidden neurons. From Table 2.6, Table 2.5 is obtained. Table 2.5 shows the improved recognition rates with different numbers of hidden neurons.

Table 2.6 shows that when the number of hidden neurons were selected in the HMM-MLPs, the recognition rates of each type of symbols with preprocessing are much more higher than the ones of the same data without preprocessing. From Table 2.6, it can be seen that the highest recognition rates were achieved when 30 and 35 hidden neurons were used in the HMM-MLP recognisers of digits and characters, respectively. The highest recognition rates are identified by the frames in Table 2.6. Tables 2.7, 2.8 and 2.9 detail recognition result of each digit, uppercase and lowercase characters, respectively. From these three tables,

Table 2.7. Recognition rates with/without preprocessing of digits based on the HMM-MLP recognisers with 30 hidden neurons

Digit	N.digits	Without Preprocessing N.correct	N.wrong	R.rate	With Preprocessing N.correct	N.wrong	R.rate
0	1154	1004	150	87.00%	1112	42	96.36%
1	676	548	128	81.07%	640	36	94.67%
2	1392	1248	144	89.66%	1376	16	98.85%
3	1340	1208	132	90.15%	1308	32	97.61%
4	1218	1096	122	89.98%	1162	56	95.40%
5	1206	1090	116	90.38%	1182	24	98.01%
6	1120	1078	42	96.25%	1106	14	98.75%
7	1280	1170	110	91.41%	1244	36	97.19%
8	1204	790	414	65.61%	1164	40	96.68%
9	1384	1228	156	88.73%	1360	24	98.27%
Total	11974	10460	1514	87.36%	11654	320	97.33%

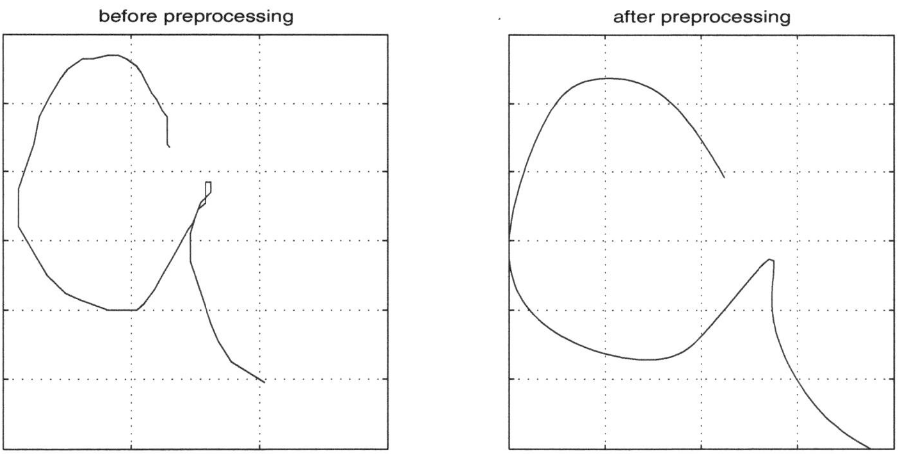

Fig. 2.9. The shapes of character 'a' before and after preprocessing

recognition rates of some special symbols without preprocessing are much lower, such as digit 8, characters D, H, K, W, Y, f, h, m, u, w, z and so on. Therefore, the recognition rates of the symbols are very much influenced by the preprocessing step. In contrast to the pre-processed symbols, the recognition rates are much higher (see on the right-hand side of each table). Table 2.10 shows the maximum recognition rates of digits and characters with preprocessing and without preprocessing, and their associative increased recognition rates, when the HMM-MLP recognisers were used.

From Tables 2.4 2.5 and 2.10, it can be seen that if the preprocessing technique was used to clear data, the increased recognition rates from the HMMs are much higher than the ones from the HMM-MLPs. As mentioned above, the features used in the HMMs are only local features. Therefore, if the local jagged points are not smoothed by the preprocessing approach, the local features are inaccurately detected, and cannot accurately describe the profile of symbols. For instance, the sharp points where the writing direction starts to change can be efficiently detected, due to many jagged points. In contrast to the local features, the global features less depend on the local information or the jagged form of symbols. For the HMM-MLP recognisers, global features were used to reduce this limitation (the bad local feature). For example, the global feature – 'number of strokes', which is obtained by counting "pen down/up", is independent on the local information (which presents jagged points); and the global features – the zone features, which are extracted by shrinking and dividing a symbol into a number of blocks, are also less dependent on the local sequence of the points. Therefore, the preprocessing is more important for local features than for the global features.

Table 2.8. Recognition rates with/without preprocessing of uppercase characters based on the HMM-MLP classifiers with 35 hidden neurons

sym-		Without Preprocessing			With Preprocessing		
bol	N.symb.	N.correct	N.wrong	R.rate	N.correct	N.wrong	R.rate
A	1170	965	205	82.15%	1078	92	92.10%
B	764	675	89	87.92%	738	26	96.60%
C	856	677	179	78.73%	816	40	95.33%
D	817	617	200	75.17%	755	62	92.42%
E	1077	998	79	91.74%	974	103	90.41%
F	688	510	178	73.67%	575	113	83.58%
G	629	435	194	68.66%	573	56	91.08%
H	771	541	230	69.78%	648	123	82.38%
I	914	636	278	69.21%	760	154	81.69%
J	596	382	214	63.58%	558	38	93.62%
K	733	420	313	56.84%	681	52	92.90%
L	656	590	66	89.63%	632	24	96.41%
M	636	433	203	67.14%	600	36	94.34%
N	783	543	240	69.02%	679	104	86.70%
O	1421	1193	228	83.72%	1277	144	89.87%
P	781	692	89	88.22%	739	42	94.62%
Q	633	511	122	80.29%	579	54	91.48%
R	924	701	223	75.53%	882	42	95.45%
S	1000	887	113	88.37%	992	8	99.20%
T	1045	954	91	91.01%	943	102	90.25%
U	781	604	177	76.97%	715	66	91.56%
V	625	491	134	78.09%	555	70	88.78%
W	707	551	156	77.54%	649	58	91.81%
X	604	392	212	64.33%	544	60	90.07%
Y	636	482	154	75.37%	602	34	94.65%
Z	539	418	121	76.94%	525	14	97.44%
Total	20786	16297	4489	78.40%	19069	1717	91.74%

Table 2.9. Recognition rates with/without preprocessing of lowercase characters based on the HMM-MLP classifiers with 35 hidden neurons

sym-		Without Preprocessing			With Preprocessing		
bol	N.symb.	N.correct	N.wrong	R.rate	N.correct	N.wrong	R.rate
a	2770	2525	245	91.15%	2544	226	91.84%
b	861	617	244	71.66%	767	94	89.07%
c	1349	1020	329	75.63%	1217	132	90.19%
d	1573	1299	274	82.61%	1491	82	94.78%
e	4551	4026	525	88.46%	4425	126	97.24%
f	1011	546	465	54.00%	857	154	84.80%
g	1128	757	371	67.14%	980	148	86.86%
h	1725	1124	601	65.14%	1596	129	92.53%
i	2141	1784	357	83.31%	2048	93	95.65%
j	625	452	173	72.31%	609	16	97.40%
k	958	552	406	57.66%	855	103	89.24%
l	2096	1531	565	73.03%	1811	285	86.41%
m	1256	958	298	76.26%	1129	127	89.92%
n	2315	1792	523	77.42%	2141	174	92.48%
o	3454	2918	536	84.47%	3201	253	92.67%
p	1358	1056	302	77.78%	1307	51	96.25%
q	845	355	490	41.96%	718	127	84.41%
r	1606	1254	352	78.09%	1477	129	91.94%
s	2112	1866	246	88.33%	2053	59	97.23%
t	2864	2514	350	87.77%	2618	246	91.40%
u	1727	775	952	44.85%	1355	372	78.18%
v	751	482	269	64.23%	620	131	82.57%
w	1043	636	407	60.95%	980	63	93.93%
x	566	290	276	51.25%	460	106	81.27%
y	1068	442	626	41.37%	932	136	87.29%
z	739	431	308	58.26%	694	45	93.96%
Total	42492	32000	10492	75.31%	38886	3606	91.51%

Table 2.10. The maximum recognition rates with/without the preprocessing step based on the HMM-MLP method

	Rec.Rate of Digits %	Rec.Rate of U.Chars.%	Rec.Rate of L.Chars.%
uppreproc.	87.36	78.4	75.31
preproc.	97.33	91.74	91.51
increased rate %	+9.97	+13.34	+16.2

We believe that most of the existing preprocessing models cannot be absolutely efficiently applied to all symbols. Therefore, the proposed approach also cannot efficiently smooth all of online handwritten symbols, or remove the hooks and noise. Although the proposed approach can successfully improve the recognition rates, it can distort the profile of some special symbols which were very badly written. Figure 2.9 shows one of the examples for the character "a". The original version of character "a" is on the left-hand side, and the other is on the right in Figure 2.9. In contrast to the version before preprocessing, the contour of the character is too smooth, and some important corners are missing. However, the proposed approach can very efficiently smooth the handwriting data, reduce noise and hooks for most of handwritten symbols. Thus, a feature extraction algorithm can efficiently extract a set of the strong discriminating features of the handwriting. Consequently the recognition rates can be improved by this preprocessing approach.

2.4 Summary

An efficient and new approach for preprocessing online handwriting recognition is presented in this study. The algorithm of the approach first removes the duplicated points and hooks of a stroke, then to filter the stroke, and finally, to normalise the online handwritten symbol. The equal-length method and the technique of detecting sharp points are applied to eliminate hooks. They also are coupled with the Cubic Spline to efficiently smooth handwritten data. In the normalisation step, the process will not deal with mathematical expressions, because the symbol's placements have to be kept for structural analysis. The evaluation of the preprocessing approach was done by visualisation and recognisers. The visualisation is done by giving a number of examples with specific symbols. The evaluation method that uses recognisers is done by the experiments. Three sets of experiments were independently carried out on digits, uppercase and lowercase characters from the benchmark UNIPEN Database. In each set of the experiment, the same classifiers are used to classify the symbols with and without preprocessing. The experiments show that this proposed preprocessing approach is efficient for online handwritten data. In addition, the discussion is given with experimental results. The comparison between the proposed approach and other existing approaches should be focused on in the future.

References

1. `http://www.vanguardsw.com/dphelp4/dph00109.htm`
2. Deepu, V., Madhvanath, S., Ramakrishnan, A.G.: Principal component analysis for online handwritten character recognition. In: ICPR 2004, pp. II: 327–330 (2004)
3. Feng, G.: Data smoothing by cubic spline filters. IEEE Trans. Signal Processing 46, 2790–2796 (1998)
4. Guerfali, W., Plamondon, R.: Normalizing and restoring on-line handwriting. Pattern Recognition 26(3), 419–431 (1993)
5. Guyon, I., Schomaker, L., Plamondon, R., Liberman, M., Janet, S.: Unipen project of on-line data exchange and recognizer benchmarks. In: Proceedings of the 12th International Conference on Pattern Recognition, ICPR 1994, pp. 29–33 (October 1994)

6. Huang, B.Q., Kechadi, M.-T.: A hybrid hmm-svm method for online handwriting symbol recognition. In: The 6th International Conference on Intelligent Systems Design and Applications, Jian, Shandong, China, pp. 887–891. IEEE Computer Society, Los Alamitos (2006)
7. Bing, Q.H., Kechadi, M.-T.: A fast feature selection model for online handwriting symbol recognition. Machine Learning and Applications. In: ICMLA 2006. 5th International Conference on 2006, pp. 251–257 (December 2006)
8. Huang, B.Q., Rashid, T., Kechadi, M.-T.: Multi-context recurrent neural network time series applications. International Journal of Computational Intelligence 3(3) (2006)
9. Chew Lim, T., Ruini, C., Peiyi, S.: Wavelet applications in segmentation of handwriting in archival documents. In: Wavelet Analysis and Its Applications: Second International Conference, Hong Kong, China, LNCS (2004)
10. Plamondon, R., Srihari, S.: On-line and off-line handwriting recognition: A comprehensive survey. IEEE PAMI 22(1), 63–84 (2000)
11. Rabiner, L.R.: A tutorial on hidden markov models and selected applications in speech recognition. Proceedings of the IEEE 77(2) (February 1989)
12. Ramsay, J.O.: Functional components of variation in handwriting. Journal of the American Statistical Association 95(449), 9–15 (2000)
13. Rashid, T., Huang, B.Q., Kechadi, M.-T., Brian, G.: Auto regressive recurrent neural network approach for electricity load forecasting. The Computational Intelligence Journal 3(1) (2006)
14. Rumelhart, D.E., McClelland, J.L.: Parallel Distributed Processing: Explorations in the Microstructure of Cognition, vol. 1 & 2. MIT Press, Cambridge (1986)
15. Tappert, C.C., Suen, C.Y., Wakahara, T.: The state of the art in online handwriting recognition. IEEE Trans. Pattern Anal. Mach. Intell. 12(8), 787–808 (1990)
16. Wink, A.M., Roerdink, J.B.T.M.: Denoising functional mr images: a comparison of wavelet denoising and gaussian smoothing. IEEE Transactions on Medical Imaging, 23(3), 374–387 (2004)

3

A Simple and Fast Term Selection Procedure for Text Clustering

Luiz G.P. Almeida[1], Ana T.R. Vasconcelos[1], and Marco A.G. Maia[2]

[1] Laboratório Nacional de Computação Científica
{lgonzaga,atrv}@lncc.br
[2] Pontifícia Universidade Catolica do Rio de Janeiro
mgrivet@cetuc.puc-rio.br

3.1 Introduction

Text clustering is a theme that is receiving considerable attention nowadays in areas such as text mining and information retrieval. A starting point for clustering methods applied to unstructured document collection is the selection of a vector-space model usually known as bag-of-words model. Documents are then described by a huge and sparse matrix which is due to the exceeding number of terms describing the set of documents. Although several techniques can be employed to reduce this number, the final figure is still high thus leading to a feature space of high dimensionality. This chapter presents a simple procedure that not only considerably reduces the dimension of the feature space and hence the processing time, but also produces clustering performance comparable or even better when confronted with the full set of terms.

The classical application of text clustering is the discovery of interrelated web pages groups as the result of a search on the WWW [22, 23, 24, 25, 26]. Text clustering is not only aimed to identify groups of similar texts from a collection but also to determine which group each one of these texts should be allocated to [28]. Unlike the process of text classification, which has the purpose of classifying texts according to a set of pre-defined classes, the text clustering has an embedded process of unsupervised learning, where text group's discovery is performed without any prior information that could link each text to some specific group [29].

As mentioned above, a starting point of almost all clustering methods applied to unstructured document's collection is the creation of a vector-space model [1, 2, 28]. The central idea is to treat their constituent words (hereafter called terms) as features and then describe each document by a vector that represents the frequency of each term's occurrence in the document. The set of all documents is then described by the so-called document-to-term matrix whose **ij**-element indicates the frequency (absolute, relative or normalized) of term **j** in the document **i**. Therefore this matrix has N_D rows and N_T columns which respectively represent the total number of documents and terms in the document collection. Needless to say, this matrix tends to be huge basically due to the fact

N. Nedjah et al. (Eds.): Intelligent Text Categorization and Clustering, SCI 164, pp. 47–64.
springerlink.com

that the total set of terms is the union of all individual document set of terms. An important feature of this representation is that the order in which the words appear in the text is not considered relevant. The text is then represented as a unordered list of terms with no semantic information [2].

Let's initially discuss the databases used for benchmark purposes that are mentioned in the literature and shortly described in table 3.1. This table presents four subsets of the Reuters text collection (R01, R02, R03 and R04) and the subset (S01) of the Medline (medical abstracts), Cranfield (aeronautical systems abstracts) and CISI (information science abstracts) from the Smart dataset[1]. We have considered the choice of these pre-classified bases as appropriate because it makes possible not only the performance's evaluation of our techniques but also allow us to compare these results with those presented in the technical literature since these databases are often used as benchmark.

Table 3.1. Original Databases Used as Benchmark

Dataset	#Docs	#Terms	Sparsity
R01	848	8537	98.90%
R02	1497	11462	99.26%
R03	1001	9154	99.16%
R04	381	5382	98.43%
S01	900	11179	99.24%

It is important to note that the number of terms above represents the total of terms that occur in each collection. In activities such as information retrieval, text classification and clustering it is common to preprocess these full term's list, with the objective of reducing the number of terms. This preprocessing [2, 28] consists of tasks such as the exclusion of words without informational value known as stop words[2], the reduction of the words to its radical and known as stemming[3], and the uppercase-lowercase conversion known as case-folding. The result of these preprocessing techniques when applied to these bases shows typically a reduction of 33% in the number of terms as described in table 3.2.

Another important feature of the document-to-term matrix is related to its sparsity. Since the list of relevant terms is the union of similar lists for each document, it is likely that some specific terms should appear only in documents of the same class. This observation does reveal (see tables 3.1 and 3.2) that the proportion of nonzero entries of this matrix rarely exceeds 1% of the total, thus indicating how severely sparse these matrices are. As it will be discussed later, this feature plays a central hole on the proposed algorithm for term selection.

[1] These datasets can be downloaded from ftp://ftp.cs.cornell.edu/pub/smart and http://www.daviddlewis.com/resources/testcollections/reuters21578/

[2] A list of common use can be found at ftp://ftp.cs.cornell.edu/pub/smart/english.stop

[3] The Algorithm Porter [44] is commonly used for stemming and it is available at http://tartarus.org/martin/PorterStemmer/

Table 3.2. Benchmark Databases after Preprocessing

Dataset	#Docs	#Terms	Sparsity
R01	848	5666	98.96%
R02	1497	7735	99.31%
R03	1001	6316	99.23%
R04	381	3572	98.51%
S01	900	7272	99.27%

The outline of this chapter is as follows: Section 2 discusses the principal clustering frameworks used today and its numerical complexity. Section 3 discusses techniques for term selection and reduction as well as the proposed algorithm for this goal. Section 4 discusses related works and how this text can be seen in the context of previous works. Section 5 presents numerical results related to the performance of our term selection method together with the techniques discussed in last section. Finally Section 6 concludes this work.

3.2 Clustering Frameworks

The purpose of any clustering algorithm is to identify the similarity of texts present in a collection and by means of some predefined criterion to determine which group each of these texts should be allocated to. When this collection is represented by a vector-space model, the similarity is computed by measuring the distance between the texts. Two measures are frequently used for the evaluation of text similarity when they are represented by vectors such as x_i e x_j: the Euclidean Distance (eq. 3.1) and the Cosine angle φ (eq. 3.2) [29] as shown below. Nevertheless we emphasize that any metric or distance can be alternatively used.

$$d(x_i, x_j) = \sqrt{(\sum_{k=1}^{N_T} |x_{ik} - x_{jk}|^2)} \tag{3.1}$$

$$\cos\varphi = \frac{< x_i, x_j >}{\| x_i \| \cdot \| x_j \|} \tag{3.2}$$

The classical clustering procedure for vector-space models is the so-called K-means algorithm [30]. This method has the following advantages: easy of implementation and low computation cost since its complexity is $O(nkl)$, where n is the number of objects, k is the number of groups and l is the number of iterations performed to achieve convergence.

The K-means algorithm promotes the partition of the set of objects $X = \{x_1, x_2, x_3, \ldots, x_n\}$, into k distinct groups, $C = \{C_1, C_2, C_3, \ldots, C_k\}$. Each group C is characterized by a central point m_c, called the centroid, which is the arithmetic mean of vectors belonging to this group and the points belonging to each group are those nearest to its centroid.

Although fast and simple to implement, this algorithm suffers from severe drawbacks such as i) it works well only for linearly separable data; ii) it is prone to local minima. Several alternative conceptual frameworks have been developed to generate better solutions for this problem.

One line of these frameworks goes towards the use of kernels in higher dimensional spaces, giving rise to algorithms such as Partitional K-means [4], Kernel K-means [5] and Weighted Kernel K-means [6] where some of the original limitations are circumvented. On the other hand, these methods can profit from techniques such as Particle Swarm Optimization (PSO) [7, 31], Genetic Algorithms (GA) [8] and others in order to search for global rather than local solutions and this hybrid approach is used to find initials centroids for the K-means algorithm. A similar strategy is another hybrid algorithm [33] that uses the K-means and Differential Evolution algorithms [32] with the same purpose of finding initial centroids.

Another line of framework is related to graph theory [9] with focuses on clustering nodes of a graph. Among graph partitioning objectives, we can mention the optimization of Ratio/Normalized Association/Cut and Kernighan-Lin parameters [6]. All these objectives can be evaluated by solving an integer-mixed Lanczos-type [18] optimization problem that involves a data affinity matrix of size $N_D \times N_D$, aimed to measure item-item similarity. Each one of its elements is a decreasing function of the distance (whose assessment is of the order N_D) between two documents in the feature space. Therefore, the assessment of the affinity matrix is a heavy numerical burden.

The affinity matrix is built so that its element (i, j) describes the similarity between the objects i e j. This similarity can be defined in various ways, but the most common is by means of the kernel RBF (Radial Basis Function) [34], also known as Kernel Gaussian:

$$R_{(i,j)} = \exp^{\frac{-d(i,j)}{2\sigma^2}} \tag{3.3}$$

where σ is a free control parameter and $d(i, j)$ is a similarity measure between objects **i** and **j**.

Finally a third line goes towards spectral algorithms that use information contained in the eigenvectors of a data affinity matrix to detect the structure present on the data [10, 11]. The numerical burden here is similar to the previous case.

Spectral algorithms were initially used for image segmentation with the purpose of clustering image points that share common characteristics such as color, brightness, contrast, etc. Nevertheless, this technique can also be used for grouping any type of data, including texts. Experiments have shown that this approach is strongly indicated to group nonlinearly separable data [35, 36].

The literature presents several different spectral algorithms [34]. In one of them [10], given a set of objects, the spectral algorithm seeks to solve the problem of dividing them into k groups, analyzing the k largest eigenvectors of the related affinity matrix. These eigenvectors tend to reflect the group structure existent in the object's collection. By forming a new matrix that is a function of a matrix whose columns are made of the row normalization of the k above mentioned largest eigenvectors and applying the K-means algorithm on the rows of this matrix, the clusters are then defined. Note that this clustering is strongly dependent on the σ parameter cited above, but [37] presents a technique to automatically determine the its optimal value.

The current literature on clustering methods reveals strong indications the the best results are obtained when using graph-based and/or spectral-based algorithms properly mixed with suitable kernels for structured data [12]. Unfortunately, most of these methods do require eigenvector's evaluation of large matrices which again, has high computational cost. Fortunately there are iterative procedures such as Lanczos and Arnoldi methods [18], which only calculate the desired number of eigenvalues, thus considerably reducing the computational time required.

3.3 The Proposed Algorithm

Last section has revealed that the most efficient clustering algorithms suffers from two bottlenecks as far as processing time is concerned. They are respectively related to the assessment of the affinity matrix and the k largest eigenvector's evaluation.

The algorithm here proposed is solely concerned with computational effort savings connected to the affinity matrix evaluation. In the general case, this evaluation requires roughly $N_D^2.N_T$ operations and the purpose here is to envisage a method that requires $N_D^2.N_M$ operations, where $N_M << N_T$. Hence we can expect computational saving´s ratio of the order N_M/N_T.

Techniques used for this purpose are in the realm of the so-called Feature Selection, which aim to reduce the number of terms that effectively contribute to the document's clustering without major impact in the classification's performance. It is important here to distinguish these techniques from those in the realm of the so-called Feature Extraction, whose central idea is to transform the whole set of features into another one with lower dimension. Examples of these techniques are Principal Components Analysis (PCA) [38], Single Value Decomposition (SVD) [39] among others.

Basically any Feature Selection algorithm has three by-products as far as Text Clustering are concerned. They are:

- the capability of mitigating the problems caused by the high dimensionality and sparsity related to similarity's assessment of documents;
- the capability of reducing the computational cost of the clustering algorithms concerning processing time and memory space;
- the capability of providing a succinct set of terms that can characterize and discriminate the clusters thus formed.

Before defining the proposed algorithm, we must first select a suitable measure for term's quality present in a specific document. The central idea here is to characterize a document not by the presence/absence of terms but instead by some meaningful quantity that expresses both term´s frequency and commonality. One possible measure of this type is the well-known parameter so-called Term Frequency Inverse Document Frequency and denoted by $TF \times IDF$ [2, 13] as expressed below:

$$TF \times IDF_{i,j} = f(t_i, D_j) = TF_{ij} \times \log(\frac{N_D}{DF_i}) \tag{3.4}$$

where t_i and D_j are respectively the term i and document j, TF_{ij} is the frequency (either absolute, relative or normalized) of term i in the document j and DF_i is the number of documents in which t_i appears.

Once defined the relative importance of terms, it is necessary to define the overall contribution of a specific term to document's discrimination. There are several ways of doing that and the most cited in the literature are: Document Frequency (DF) [15], Term Strength (TS) [40], Term Variance Quality (TVQ) [17], Term Variance (TV) [2], Entropy (En) [16] and Term Contribution (TC) [2, 13]. These measures were compared in two articles. In the first [2] measures DF, TVQ, TV and TC are compared and the best results are obtained with TC and TV, with a little advantage for TC. In the second study [3] DF, En, TS and TC are compared and the best results are obtained with the TS and TC. The authors concluded that TC is the most advantageous in terms of lower computational cost.

Based upon these results, we decided to use the TC in the proposed algorithm, although we acknowledge that any other measure can be used since specific problems may present better results with different measures. The definition of TC for term **k** is given by [2]:

$$TC(t_k) = \sum_{i=1}^{N} \sum_{j=1,\ j \neq i}^{N} w_{(t_k, d_i)} \times w_{(t_k, d_j)} \tag{3.5}$$

where $w_{(t,d_i)}$ is given by:

$$w_{(t_k, d_i)} = \frac{TF \times IDF_{k,i}}{\sqrt[2]{\sum_n (TF \times IDF_{n,i})^2}} \tag{3.6}$$

The problem we are facing here can be stated as follows: how to select terms in such a way that:

- they cover all documents;
- they are "good" for clustering purposes;
- their number is much smaller than N_T.

The formal statement of an optimization problem that can solve term selection as stated above is in the realm of Combinatorial Optimization and for this reason, it is hard and numerically expensive to solve. One possible alternative is to investigate ad-hoc algorithms based upon some reasonable heuristics.

One of such naive procedures consists of sorting the terms in decreasing order of their Term Contribution TC and then selecting at least the first terms that cover the whole set of documents. This method does not necessarily give the best solution and perhaps a less greedy technique can provide better results. For future references, we will call this method as GREEDY ALGORITHM.

A little and richer variation of this scheme can be envisaged and its performance can be investigated. Let L be a positive integer and suppose that we sort the terms in decreasing order of their Term Contribution (TC) (or another measure). Then we select the top L terms with highest TC's. Since the document-to-term matrix A is very sparse, it is unlikely that these L terms cover all documents. Then we remove from A all covered documents, re-evaluate the Term Contribution of the remaining terms and sort them again in decreasing order. As before, we select the top L terms and if there are uncovered documents, we repeat this procedure until all documents are covered. Formally speaking, this term selection procedure can be described as follows:

1. Let A be the document-to-term matrix associated to a set of N_D documents and N_T terms
2. Evaluate for each term of A its Term Contribution TC as in equations 3.4 and 3.5
3. Find the top L terms with highest TC, putting them in list S.
4. Remove from A all documents (rows) covered by S and also remove from A all terms (columns) in S.
5. If A has no rows, go to Step 6. Otherwise, go to Step 2
6. Stop. S is the list of selected terms.

This algorithm is extremely simple to implement and it has the parameter L that can be adjusted for the best performance. We will call this method as L-COVERING METHOD.

Notice that the greedy algorithm can be seen as a particular case of the above method, by taking $L = \infty$. From a practical point of view, it is reasonable to accept that as L increases, the performance of this method should get closer to its greedy version. An important feature of this algorithm is the freedom of choice of the parameter L, because it can be tuned to produce best results for specific applications.

3.4 Related Works

According to [13], feature selection methods have been successfully applied to text categorization but seldom applied to text clustering due to the unavailability of class label information. This is no longer true and today several databases focusing on different kind of documents are freely available.

The vast majority of papers on feature selection work on the principle of the greedy algorithm previously described or in some small variation of it. What distinguish them is the criterion for term's quality and in most of them the number of terms is arbitrarily selected, based solely upon on computational effort. For instance, Yang [14] advocates a term strength measure, Yang and Pedersen [15] defends a term's information gain, Dash and Liu [16] defines an entropy-based ranking, Dhillon [17] introduces the term variance quality and term profile quality while Liu and Cheng [13] invented the term contribution criteria expressed in 3.4. In the approach proposed in [41], which also uses the entropy to select the terms, the process is more sophisticated, since there is a concern that all documents are represented after term selection but this algorithm requires prior determination of the minimum number of terms.

The results presented in all these works are clustering-related performance's measures expressed as a function of the top term's fraction selected according to the advocated criterion.

We believe that, due to the high level of the document-to-term matrix's sparsity, it is possible that some bias can occur towards those documents on the top ranking of the criterion classification when the greedy algorithm is deployed and hence it is likely that L-covering method can reduce such a bias. Nevertheless we must take into consideration that greedy behavior is a particular case of the L-covering end hence if better results are reached with the former, so be it. What in our opinion cannot be ignored is the possibility the L-covering can outperform its greedy version in most cases of interest.

3.5 Numerical Results

Before presenting performance results, it is important to discuss some features of the text databases used in all experiments in this work. Two collections of pre-classified documents were used: the Smart and Reuters21578 bases. From these two datasets, five new datasets were produced: the base S01 composed of 3 categories (cisi, cran and med) derived from Smart base and the bases R01, R02, R03 and R04, derived from Reuters21578 base, respectively composed of 11, 5, 7 and 10 categories. Next, each of them were populated by a random sampling the corresponding full collection. The number of documents, terms and the sparsity of each base are shown in tables 3.1 and 3.2.

When the Greedy and L-Covering methods was applied to these databases, the number of remaining terms and sparsity for some values of L are respectively presented on Tables 3.3 and 3.4, 3.5, 3.6, 3.7, 3.8.

When Greedy algorithm is deployed, the term's reduction has ranged from 0.78% to 2.93% while in all cases with L-Coverage Algorithm, this reduction has ranged from 1% to 1.6%. These figures show that processing time for the affinity

Table 3.3. Greedy method performance

		Greedy method		
Base	Terms	Execution time	Terms	Percent
R01	5666	0.92	58	1.02%
R02	7735	1.58	169	2.18%
R03	6316	1.16	185	2.93%
R04	3572	0.46	28	0.78%
S01	7272	1.20	99	1.36%

Table 3.4. Base R01, 5666 terms

L	Execution time	Selected terms	Percent	Sparsity
1	3.21	38	0.67%	93.89%
2	1.67	42	0.74%	93.57%
3	1.35	57	1.01%	94.00%
4	1.05	60	1.06%	93.43%
5	0.68	45	0.79%	91.59%

Table 3.5. Base R02, 7735 terms

L	Execution time	Selected terms	Percent	Sparsity
1	12.39	83	1.07%	93.91%
2	4.57	68	0.88%	92.98%
3	3.21	66	0.85%	92.00%
4	3.00	80	1.03%	93.85%
5	2.39	95	1.23%	93.89%

Table 3.6. Base R03, 6316 terms

L	Execution time	Selected terms	Percent	Sparsity
1	6.27	55	0.87%	93.83%
2	3.25	54	0.85%	93.55%
3	1.92	60	0.95%	93.48%
4	1.52	56	0.89%	92.90%
5	1.19	60	0.95%	93.04%

Table 3.7. Base R04, 3572 terms

L	Execution time	Selected terms	Percent	Sparsity
1	1.37	33	0.92%	92.01%
2	0.76	40	1.12%	91.71%
3	0.45	33	0.92%	91.31%
4	0.35	36	1.01%	91.48%
5	0.24	25	0.70%	89.04%

Table 3.8. Base S01, 7272 terms

L	Execution time	Selected terms	Percent	Sparsity
1	16.37	116	1.60%	96.63%
2	7.63	114	1.57%	96.58%
3	4.77	111	1.53%	96.00%
4	3.52	108	1.49%	96.23%
5	2.46	100	1.38%	95.97%

Table 3.9. Base R01, 848 documents, 11 categories

Case			AA	RS	FM	t
K-means	FS	5666	69.19%	79.79%	53.61%	0.00
Spectral	FS	5666	53.91%	77.04%	49.62%	0.00
K-means	F20	1133	69.92%	80.09%	54.61%	4.85
Spectral	F20	1133	56.13%	77.06%	46.77%	1.02
K-means	F15	850	69.88%	80.06%	54.47%	3.75
Spectral	F15	850	56.50%	77.10%	46.64%	1.00
K-means	F10	567	69.74%	80.05%	54.47%	2.50
Spectral	F10	567	60.75%	77.95%	48.59%	1.00
K-means	F5	283	69.87%	80.25%	55.12%	1.28
Spectral	F5	283	63.66%	78.36%	49.45%	0.99
K-means	F3	170	68.15%	80.00%	54.54%	0.88
Spectral	F3	170	64.46%	78.70%	50.57%	1.23
K-means	GD	58	64.20%	78.68%	50.55%	1.02
Spectral	GD	58	61.36%	79.31%	54.08%	2.10
K-means	L5	45	68.75%	80.31%	55.53%	0.38
Spectral	L5	45	63.09%	78.07%	48.39%	1.12
K-means	L4	60	73.03%	81.56%	59.00%	0.38
Spectral	L4	60	65.94%	78.98%	51.32%	0.93
K-means	L3	57	73.96%	81.80%	59.69%	0.37
Spectral	L3	57	70.45%	80.43%	55.70%	0.90
K-means	L2	42	78.19%	83.75%	65.08%	0.36
Spectral	L2	42	75.08%	82.24%	60.99%	0.91
K-means	L1	38	79.90%	84.58%	67.30%	0.29
Spectral	L1	38	75.09%	82.33%	61.21%	0.90

matrix evaluation can be in the range of one to two orders of magnitude smaller when these feature's selection method are deployed.

In order to assess clustering performance under different feature's selection methods, three quality measures were selected for this purpose: the Average Accuracy (AA) [2], the Rand statistics (RS) [42] and Folks and Mallow index (FM) [43] (see [2] for details). These measures are external or supervised, that is, depend upon the knowledge of previous classification.

The performance results for the five databases processed by plain K-means and Spectral Clustering algorithms are presented in Tables 3.9, 3.10, 3.11, 3.12 and 3.13. In these tables:

- FS stands for FULLSET of terms;
- F $<$ *number* $>$ stands for percent of terms[4] (indicated by $<$ *number* $>$) arbitrarily selected from FULLSET;
- GD stands for GREEDY set of terms;
- L$<$ *number* $>$ stands for L-COVERING ($<$ *number* $>$ is L) set of terms;
- last column indicates the processing time in seconds.

[4] Terms are sorted in decreasing order of their Term Contribution.

Table 3.10. Base R02, 1497 documents, 5 categories

Case			AA	RS	FM	t
K-means	FS	7735	67.73%	79.79%	53.61%	261.19
Spectral	FS	7735	48.70%	81.65%	60.77%	8.50
K-means	F20	1547	64.98%	88.55%	74.09%	9.79
Spectral	F20	1547	46.45%	79.55%	59.90%	3.88
K-means	F15	1160	68.94%	89.85%	76.49%	7.16
Spectral	F15	1160	50.79%	82.50%	62.93%	3.88
K-means	F10	774	71.68%	90.77%	78.39%	4.04
Spectral	F10	774	52.41%	83.41%	64.12%	3.90
K-means	F5	387	72.91%	91.20%	79.19%	2.26
Spectral	F5	387	55.75%	84.93%	67.22%	3.90
K-means	F3	232	78.66%	93.06%	83.05%	1.27
Spectral	F3	232	58.11%	85.94%	68.69%	3.90
K-means	GD	169	76.73%	92.41%	81.64%	2.21
Spectral	GD	169	65.70%	88.72%	72.99%	8.16
K-means	L5	95	76.72%	92.33%	80.95%	0.56
Spectral	L5	95	72.00%	90.75%	77.10%	3.86
K-means	L4	80	79.01%	93.09%	82.73%	0.62
Spectral	L4	80	67.51%	89.16%	73.24%	3.80
K-means	L3	66	77.68%	92.64%	81.61%	0.53
Spectral	L3	66	67.63%	89.23%	73.49%	3.84
K-means	L2	68	75.61%	91.93%	79.85%	0.61
Spectral	L2	68	59.80%	86.38%	67.01%	3.81
K-means	L1	83	70.77%	90.26%	75.77%	0.65
Spectral	L1	83	67.70%	89.20%	73.26%	3.85

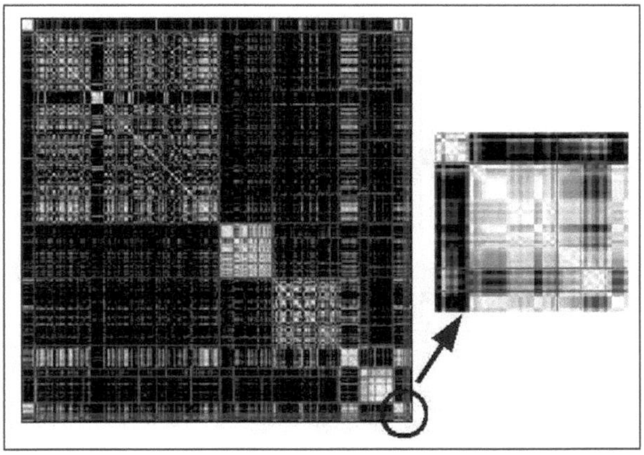

Fig. 3.1. Affinity matrix for the R01 dataset. 5666 terms.

Table 3.11. Base R03, 1001 documents, 7 categories

Case			AA	RS	FM	t
K-means	FS	6316	79.58%	94.04%	81.50%	120.99
Spectral	FS	6316	59.26%	88.36%	66.87%	2.95
K-means	F20	1263	81.02%	94.57%	83.28%	4.85
Spectral	F20	1263	56.30%	87.34%	66.83%	1.37
K-means	F15	947	81.73%	94.76%	83.81%	3.38
Spectral	F15	947	73.93%	92.83%	78.44%	1.36
K-means	F10	632	82.52%	95.00%	84.60%	2.27
Spectral	F10	632	76.25%	93.57%	80.74%	1.36
K-means	F5	316	82.71%	95.02%	84.64%	1.14
Spectral	F5	316	73.77%	92.50%	76.89%	1.35
K-means	F3	189	82.51%	94.98%	84.56%	0.71
Spectral	F3	189	74.30%	92.66%	77.42%	1.35
K-means	GD	185	80.86%	94.52%	83.18%	0.00
Spectral	GD	185	76.40%	93.65%	81.06%	2.99
K-means	L5	60	82.44%	95.06%	84.84%	0.30
Spectral	L5	60	77.97%	93.86%	81.49%	1.29
K-means	L4	56	81.42%	94.74%	83.81%	0.28
Spectral	L4	56	79.48%	94.25%	82.40%	1.31
K-means	L3	60	83.03%	95.25%	85.45%	0.32
Spectral	L3	60	79.39%	94.21%	82.40%	1.30
K-means	L2	54	80.94%	94.49%	82.87%	0.29
Spectral	L2	54	78.84%	93.94%	81.26%	1.31
K-means	L1	55	80.91%	94.52%	83.01%	0.28
Spectral	L1	55	75.84%	92.91%	77.85%	1.31

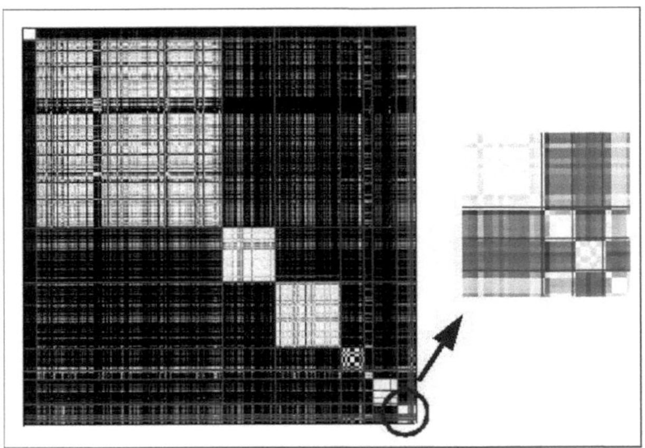

Fig. 3.2. Affinity matrix for the R01 dataset. 38 terms. (1-covering).

Table 3.12. Base R04, 381 documents, 10 categories

Case			AA	RS	FM	t
K-means	FS	3572	69.68%	94.33%	72.93%	16.78
Spectral	FS	3572	74.68%	95.41%	78.17%	0.48
K-means	F20	714	69.40%	94.30%	72.88%	1.36
Spectral	F20	714	80.49%	96.37%	82.38%	0.24
K-means	F15	536	70.17%	94.48%	73.75%	1.01
Spectral	F15	536	78.01%	95.85%	79.83%	0.23
K-means	F10	357	71.01%	94.63%	74.47%	0.65
Spectral	F10	357	80.19%	96.21%	81.46%	0.23
K-means	F5	179	70.93%	94.64%	74.61%	0.31
Spectral	F5	179	70.27%	94.44%	73.27%	0.33
K-means	F3	107	71.82%	94.86%	75.75%	0.22
Spectral	F3	107	68.25%	94.05%	71.59%	0.22
K-means	GD	28	66.73%	93.89%	71.64%	0.22
Spectral	GD	28	73.72%	95.00%	75.69%	0.44
K-means	L5	25	59.85%	92.36%	63.85%	0.14
Spectral	L5	25	59.98%	92.39%	63.89%	0.20
K-means	L4	36	65.38%	93.56%	69.77%	0.13
Spectral	L4	36	62.01%	92.77%	65.31%	0.20
K-means	L3	33	43.50%	88.89%	47.64%	0.14
Spectral	L3	33	41.61%	88.53%	44.91%	0.20
K-means	L2	40	55.61%	91.52%	60.64%	0.16
Spectral	L2	40	57.78%	91.89%	60.88%	0.20
K-means	L1	33	60.21%	92.53%	65.39%	0.13
Spectral	L1	33	58.44%	92.12%	63.42%	0.20

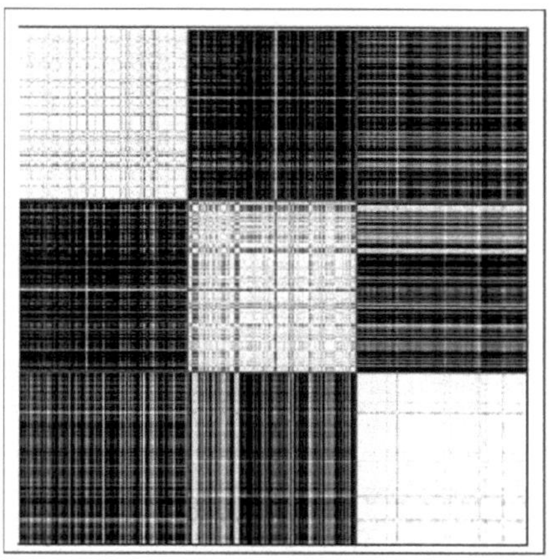

Fig. 3.3. Affinity matrix for the S01 dataset. 7272 terms.

Table 3.13. Base S01, 900 documents, 3 categories

Case			AA	RS	FM	t
K-means	FS	7272	96.70%	98.53%	97.79%	94.83
Spectral	FS	7272	81.78%	91.71%	87.66%	2.05
K-means	F20	1454	97.02%	98.67%	98.01%	3.14
Spectral	F20	1454	93.52%	97.09%	95.62%	0.94
K-means	F15	1091	97.02%	98.67%	98.05%	2.31
Spectral	F15	1091	95.43%	97.95%	96.92%	0.94
K-means	F10	727	96.90%	98.62%	97.92%	1.53
Spectral	F10	727	94.79%	97.66%	96.48%	0.95
K-means	F5	364	92.88%	96.80%	95.18%	0.79
Spectral	F5	364	93.93%	96.37%	94.55%	0.94
K-means	F3	218	91.32%	96.09%	94.13%	0.57
Spectral	F3	218	89.47%	95.24%	92.85%	0.94
K-means	GD	99	82.04%	91.72%	87.60%	0.74
Spectral	GD	99	75.76%	88.69%	83.14%	2.67
K-means	L5	100	69.20%	85.14%	78.25%	0.44
Spectral	L5	100	57.97%	79.28%	69.85%	0.97
K-means	L4	108	69.64%	85.46%	78.51%	0.45
Spectral	L4	108	43.59%	69.59%	60.45%	0.98
K-means	L3	111	58.84%	79.35%	70.75%	0.43
Spectral	L3	111	52.26%	75.89%	65.20%	0.98
K-means	L2	114	63.49%	82.09%	73.63%	0.45
Spectral	L2	114	52.56%	76.00%	66.74%	0.97
K-means	L1	116	66.92%	83.97%	76.45%	0.48
Spectral	L1	116	57.86%	79.21%	69.68%	0.97

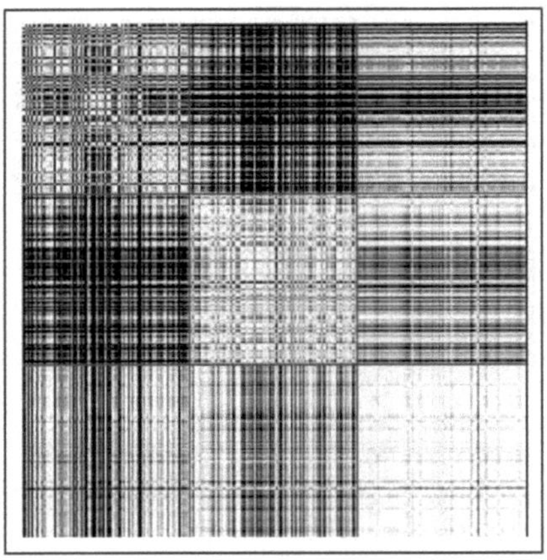

Fig. 3.4. Affinity matrix for the S01 dataset. 116 terms. (1-covering).

From these tables one can conclude that, except for the cases corresponding to R04 e S01 databases, the L-COVERING algorithm outperformed the GREEDY algorithm with fewer terms in all cases. This seems to indicate that the proposed term selection algorithm, although simple, can be employed when processing time is an asset.

We think that would be desirable at this point to discuss the intrinsic quality of the databases used for performance assessment. The simplest way of doing this is by means of the visualization of the document's similarity. An image of size $N_T \times N_T$ can be produced such that pixel at row i and column j is gray-scaled in such a way that null and maximum document's similarities are respectively represented by black and white. Ideally, this figure is made by near white blocks in the diagonal on a near black background where the block sizes correspond to cluster sizes.

As an example, figure 3.1 shows the corresponding image for the untreated R01 database. From there we can see several characteristics of this collection such as a) it is made of clusters with huge variation in size b) clusters 1 and 3 have some documents very dissimilar to other cluster members c) clusters 9, 10 and 11 are very confusing (see zoom). All these seem to indicate that clustering performance will not be very high because the dataset itself is composed by documents whose vector representation does not exhibit high intra-class/low inter-class similarities.

Take for instance the same database when the L-Covering algorithm is used, thus resulting in a database of 38 terms. Figure 3.2 shows the resulting image where we can easily see that most of the bad features above mentioned are severely mitigated.

Now we are in a position to explain why the results were not so good when database S01 is under test. Figures 3.3 and 3.4 respectively show the corresponding image for the fullset of 7272 terms and when L-covering algorithm is used, which reduces to 116 terms.

In the case of S01 database a further investigation is required, either to assess performance under different metrics or to employ some additional or alternative process for feature selection.

3.6 Conclusion

The work here discussed presents a simple procedure for term selection in the context of text clustering. Different from those reported in the literature, the selected terms are not the best ones when a specific rank criterion is deployed to assess term's quality. This algorithm takes into account the severe sparsity of the document-to-term matrix an introduces a control parameter that is useful to tune its application on different databases.

The two discussed algorithms for selection of features, namely Greedy and L-Coverage, both reduce the number of terms to about 1.3% of the amount of initial terms, promoting substantial reduction in the size of the document-term matrix. When this reduced matrix is deployed, the measured statistics clearly indicate an impact in the performance. For datasets such as R01, R02 and R03 which share a common vocabulary, the proposed feature's selection algorithm has a positive impact on the results thus showing that improvement can be obtained even with a small fraction of terms. Nevertheless, for those datasets such as S01, which is made of three groups of different vocabularies, the impact of the proposed algorithm was negative, thus indicating that further studies are required.

The results so far obtained are encouraging because they have shown a small increase in the clustering performance with an extremely large reduction on the number of terms, revealing its adequacy for those cases where processing time is the primary goal.

References

1. Salton, G., McGill, M.J.: Introduction to Modern Retrieval. McGraw-Hill Book Company, New York (1983)
2. Liu, L., et al.: A comparative study on unsupervised feature selection methods for text clustering. In: Proceeding of Natural Language Processing and Knowledge Engineering 2005, October (November 2005)
3. Duda, R.O., Hart, P.E.: Pattern Classification, 2nd edn. John Wiley and Sons, Chichester (2001)
4. Dhillon, I.S., Modha, D.S.: Concept Decompositions For Large Sparse Text Data Using Clustering. Machine Learning 42(1), 143–175 (2001)
5. T. Gïtner, A Survey of Kernels for Structured Data. ACM SIGKDD Explorations Newsletter 5(1), (July 2003)
6. Dhillon, I.S., Guan, Y., Kulis, B.: Kernel k-means, Spectral Clustering and Normalized Cuts. In: Int. Conf. on Knowledge Discovery, KDD 2005, Seattle, USA, pp.551–556 (2005)
7. Chen, C.Y., Ye, F.: Particle Swarm Optimization Algorithm and Its Application to Clustering Analysis. In: Proc. of the 2204 IEEE Int. Conf. on Networking, Sensing & Control, Taiwan (2004)
8. Tseng, L.Y., Shiueng, B.Y.: A genetic approach to the automatic clustering problem. Pattern Recognition 34(2), 415–424 (2001)
9. Chung, F.R.K.: Spectral Graph Theory, Amer. Math. Society Press (1997)
10. Ng, A.Y., Jordan, M.I., Weiss, Y.: On Spectral Clustering: Analysis and an Algorithm. In: Proc. Neural Info. Processing Systems (NIPS 14) (2002)
11. Kamvar, S.D., Klein, D., Manning, C.D.: Spectral Learning. In: Int. Joint Conf. on Artificial Intelligence, IJCAI 2003 (2003)
12. Karatzoglou, A., Feinerer, I.: Text Clustering with String Kernels in R, Dept. Statistics and Mathematics, Wirtschaftsuniversitat Wien, Report 34 (May 2006)
13. Liu, T., Liu, S., Chen, Z., Ma, W.Y.: An Evaluation on Feature Selection for Text Clustering. In: Proceeding of the Twentieth International Conference on Machine Learning (ICML 2003), Washington DC (2003)
14. Yang, Y.: Noise Reduction in a Statistical Approach to Text Categorization. In: Proc. Of SIGIR9k5, pp.256–263
15. Yang, Y., Pedersen, J.O.: A Comparative Study of Feature Selection in Text Categorization. In: Proc. Of ICML 1997, pp. 412–420 (1997)
16. Dash, M., Liu, H.: Feature Selection for Clustering. In: Proc. Of PAKDD 2000, pp. 110–121

17. Dhillon, I., Kogan, J., Nicholas, C.: Feature Selection and Document Clustering. In: Berry, M. (ed.) A Comprehensive Survey of Text Mining. LNCS. Springer, Heidelberg (2002)
18. Golub, G., Van Loan, C.: Matrix Computations, 3rd edn. John Hopkins University Press (1996)
19. Douglas, R., Cutting, D.R., Karger, J.O., Pedersen, J.W., Tukey, S.G.: A Cluster-based Approach to Browsing Large Document Collections. In: Proceedings of the 15th Annual International ACM SIGIR Conference on Research and Development in Information Retrieval, pp. 318–329 (1992)
20. Rijsbergen, C.J.: Information Retrieval, 2nd edn. Butterworths (1979)
21. Zamir, O., Etzioni, O.: Grouper: A dynamic clustering interface to web search results. Computer Networks (Amsterdam, Netherlands) 31(11-16), 1361–1374 (1999)
22. Weiss, D., Osiński, S.: Carrot2: Design of a flexible and efficient web information retrieval framework. In: Szczepaniak, P.S., Kacprzyk, J., Niewiadomski, A. (eds.) AWIC 2005. LNCS (LNAI), vol. 3528, pp. 439–444. Springer, Heidelberg (2005)
23. Tanabe, L., Scherf, U., Smith, L.H., Leel, J.K., Hunter, L., Weinstein, J.N.: Medminer: An internet text-mining tool for biomedical information, with application to gene expression profiling. BioTechniques 27, 1210–1217 (1999)
24. Yamamoto, Y., Takagi, T.: Biomedical knowledge navigation by literature clustering. Journal of Biomedical Informatics 40, 114–130 (2007)
25. Steinbach, M., Karypis, G., Kumar, V.: A comparison of document clustering techniques. In: KDD Workshop on Text Mining (2000)
26. Iliopoulos, I., Enright, A.J., Ouzounis, C.A.: Textquest: Document clustering of medline abstracts for concept discovery in molecular biology. In: Pacific Symposium on Biocomputing, vol. 6, pp. 384–395 (2001)
27. Russell, S., Norvig, P.: Artificial Intelligence: A Modern Approach, 2nd edn. Prentice Hall, Englewood Cliffs
28. Hotho, A., Nurnberger, A., Paaï, G.: A Brief Survey of Text Mining. LDV Forum - GLDV Journal for Computational Linguistics and Language Technology 20(1), 1962 (2005)
29. Kaufman, L., Rousseeuw, P.J.: Finding Groups in Data: An Introduction to Cluster Analysis. John Wiley Sons Inc., Chichester (1990)
30. MacQueen, J.B.: Some methods for classification and analysis of multivariate observations. In: 5th Berkeley Symp. Math Statist. Prob., vol. 1, pp. 281–297 (1967)
31. Cui, X., Potok, T.E.: Document clustering analysis based on hybrid pso + k-means algorithm. Journal of Computer Sciences. Special Issue: 2733 (2005)
32. Storn, R., Price, K.: Differential evolution - a simple and efficient heuristic for global optimation over continuos spaces. Jounal of Global Optimization 11, 341–359 (1997)
33. Abraham, A., Das, S., Konar, A.: Document clustering using differential evolution. In: IEEE Congress on Evolutionary Computation CEC, pp. 1784–1791 (July 2006)
34. Weiss, Y.: Segmentation using eigenvectors: a unifying view. In: Seventh International Conference on Computer Vision (ICCV 1999), 2 (1999)
35. Campbell, C.: An introduction to kernel methods. In: Howlett, R.J., Jain, L.C. (eds.) Radial Basis Function Networks: Design and Applications. Springer, Heidelberg (2000)
36. De Bie, T., Cristianini, N., Rosipal, R.: Eigenproblems in Pattern Recognition. In: Bayro-Corrochano, E. (ed.) Handbook of Geometric Computing: Applications in Pattern Recognition, Computer Vision, Neuralcomputing, and Robotics, pp. 129–170. Springer, Heidelberg (2005)

37. Zelnik-Manor, L., Perona, P.: Self-tuning spectral clustering. Advances in Neural Information Processing Systems (2004)
38. Jolliffe, I.T.: Principal component analysis. Series in Statistics. Springer, Heidelberg (1986)
39. Wall, M.E., Rechtsteiner, A. Rocha, L.M.: Singular value decomposition and principal component analysis. ArXiv Physics e-prints (August 2002)
40. John Wilbur, W.: e Karl Sirotkin. The automatic identification of stop words. Journal of Information Science 18(1), 45–55 (1992)
41. Borgelt, C., Nurnberger, A.: Experiments in document clustering using cluster specific term weights. In: Proceedings Workshop Machine Learning and Interaction for Text-based Information Retrieval (2004)
42. Rand, W.M.: Objective criteria for the evaluation of clustering methods. Journal of the American Statistical Association 66, 846–850 (1971)
43. Fowlkes, E.B., Mallows, C.L.: A method of comparing two hierarchical clusterings. Journal of the American Statistical Association 78, 553–569 (1983)
44. Porter, M.F.: An algorithm for suffix stripping. Program 14(3), 130–137 (1980)

4

Bilingual Search Engine and Tutoring System Augmented with Query Expansion

Nuttanart Facundes[1], Ratchaneekorn Theva-aksorn[2], and Booncharoen Sirinaovakul[3]

King Mongkut's University of Technology Thonburi 126 Pracha-uthit Rd., Bangmod, Thungkru, Bangkok 10140 Thailand
nuttanart@cpe.eng.kmutt.ac.th[1], ratchaneekorn.t@hcwml.com[2], boon@kmutt.ac.th[3]

Nowadays, distributed information in the form of linguistic resources is accessible throughout the internet. Search engines have become crucial for information retrieval since users need to find and retrieve relevant information in various languages and forms. The objective of this paper is to investigate the application of query expansion technique to improve cross-language information retrieval in English and Thai as well as the potential to apply the technique to other intelligent systems such as tutoring systems. As the method of evaluation of query expansion, we investigate whether the expanded terms are useful for the search.

4.1 Introduction

Query expansion is the process of a search engine adding search terms to a user's weighted search. Query Expansion techniques can be classified into three categories [1]: manual, interactive and automatic. Manual query expansion refers to techniques that a user may employ to modify the query without the aid from the system. Interactive Query Expansion means the user has some interaction with the system in the query expansion process. Automatic Query Expansion refers to techniques that modify a query *without* user control. In this work, we develop automatic query expansion system employing concept-based method [2]. In this method, query expansion means to acquire the expanded search terms from a dictionary or thesaurus. Gaining from this knowledge, query evaluation is no longer restrained to query terms submitted by users but may also embody synonymous or semantically related terms. Effective systems for improving effectiveness in information retrieval systems by query expansion technique have been available for several years. Brisebois [3] employs query expansion process uses human-defined knowledge in order to uncover indirect connections between query terms and document content by using WordNet system which provides semantic relations among terms. However, it should be noted that newly expanded terms are sometimes not as reliable as the initial terms obtained from users. An

N. Nedjah et al. (Eds.): Intelligent Text Categorization and Clustering, SCI 164, pp. 65–79.
springerlink.com

appropriate weighting scheme allows a smooth integration of these related terms by reducing their influence over the query. Accordingly, all terms recovered from the thesaurus will be given some weight expressing their similarity to the initial terms. These weights are comprised between 0 and 1 and a weight of 1 is assigned to original query terms.

4.2 Proposed Work

Our query expansion system is composed of two parts: search capabilities and browse capabilities. Searching capabilities consist of four components: Operators specification, Query expansion, Query translation and Similarity measurement.

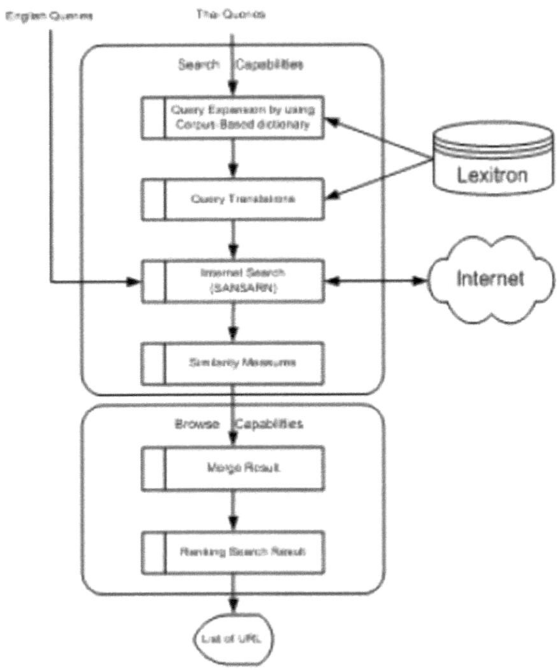

Fig. 4.1. Overview of Query Expansion System

Figure 1 shows the overview of our query expansion system. Not that we employ two existing resources: 1) LEXiTRON, an electronic, bi-directional Thai-English dictionary [4], and 2) SANSARN, an existing bilingual (English and Thai) search engine. These two resources are developed by National Electronic and Computer Technology Center (NECTEC), Thailand. Next, each module in search capabilities and browse capabilities will be discussed.

4.2.1 Search Capabilities

Operator Specification

Operator specification allows a user to logically relate multiple concepts together to define what information is needed. Our system supports queries with various combinations of Boolean operators AND, OR and NOT such as the query "mammal AND biped NOT human". Operators are processed from left to right.

Query Expansion

The query expansion component employs the dictionary-based technique using part-of-speech information and semantic relations. We use LEXiTRON for this expansion process. When a user submits a query to system, the query is expanded by adding the terms which are synonyms and related words. However, since a word can have many senses, the appropriate level of expansion must be determined. At first level expanding, we retain every possible sense, i.e. every possible part-of-speech. In the second level, we perform a part-of-speech disambiguation. If the 1st-level expanded words include more than one part-of-speech, only words that have the same part-of-speech will be selected to the 2nd-level expanded word. Proper weight calculation is then performed on the resulting words from the 2nd-level.

Query Translation

The queries are translated using the same dictionary as in the previous step. We adopt a term-by-term translation method since it is significantly less costly than other techniques. In the translation process, we also have to remove English stop words including a, an, and, at, be, by, do, for, in, of, on, or, the, to, with. The translation process is shown in the following figure.

Similarity Measurement

The similarity measurement which tells how alike two documents are, or how alike a document and a query are. For each query term q_i and its expanded set

Table 4.1. Examples of query term processing

Search Statement	System Operation
W_1 OR W_2 NOT W_3	Select all item concerning W_1 or W_2 that do not concern W_3
W_1 OR W_2 AND W_3	Select all item concerning W_1 or W_2 and items concerning W_3
W_1 NOT W_2 AND W_3	Select all item concerning W_1 but do not concern W_2 and items concerning W_3

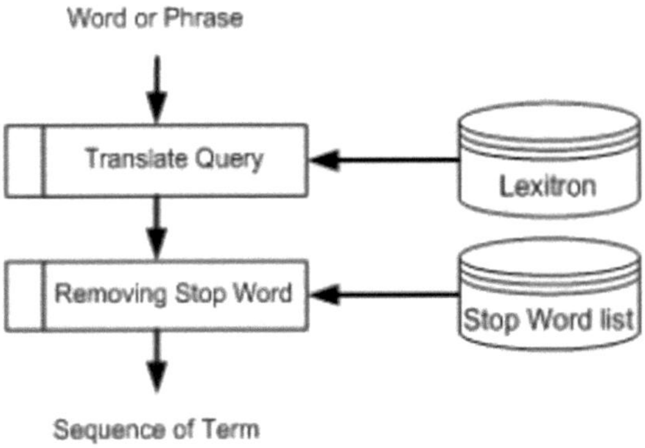

Fig. 4.2. Query Translation Process

q_{i1}, q_{i2}, ..., q_{in}. Each expanded query term retrieved the same r number of documents. Therefore, we will acquire total rn documents from this step. From rn documents, we select only top r ranking documents and return to user. To re-ranking the similarity for each document, we use the scalar vector product for the weighted retrieval algorithm. For this algorithm, the query and documents is represented by the scalar vector. This algorithms accepts a weighted query $q_i = a_{i1}, a_{i2}, \ldots, a_{in}$. Each document d_j return from SANSARN is thought to be represented by a vector $d_j = b_{j1}, b_{ji2}, \ldots, b_{jn}$. Here, the a_{in}'s and b_{jn}'s imply the weight of the expanded term qin calculated from the expanded process and calculated from the document d_j, respectively. The value i, n, and j signify the number of query, the number of expanded query and number of document retrieved from q_{ik}. Please note that as our system is concentrated on the expansion method, thus, only expanded query term is presented in the scalar vector as explain above, n is the number of expanded query term.

Now, each b_{jn} is calculated from the term frequency of that term in the document:

$$b_{ij} = \frac{f(q_{ik,d_j})}{1 + tf(q_{ik}, d_j)}$$

where: $tf(q_{ik},d_j)$ = the term frequency of the expanded term k in the document d_j

The retrieval function used by the weighted retrieval algorithm is the scalar vector product. For each expanded term, a **Retrieval Status Value** (RSV) for each document is calculated by comparing it to the query as shown below:

$$RSV(q_i, d_j) = a_{ik} \cdot b_{jk}$$

where: j = document j
a_{ik} = weight of expanded term k^{th} of query term i

Moreover, the logical OR relation is incorporated into our search capabilities. This is because, intuitively, when a query term is expanded with another term, the latter should be considered as an alternative expression of the former. Accordingly, a vector representation that integrates the logical OR relation is suggested. As discussed in [5], there are many ways to integrate the OR operator. However, we choose the following straightforward formula for OR:

$$RSV(A\_OR\_B) = RSV(A) + RSV(B)$$

We apply the above formula for our expanded term. Since we have n expanded terms for each query term, the RSV of that query term is equal to the sum of all RSV as shown below:

$$\begin{aligned} RSV(q_i, d_j) &= RSV(q_{i1}, d_i) + RSV(q_{i2}, d_i) + \cdots + RSV(q_{in}, d_i) \\ &= a_{i1} \times b_{j1} + a_{i2} \times b_{j2} + \cdots + a_{in} \times b_{jn} \\ RSV(q_i, d_j) &= \sum_{k=1}^{n} a_{ik} \cdot b_{jk} \end{aligned}$$

where: n = number of expanded term for query term i
j = document j
a_{ik} = weight of expanded term k^{th} of query term i

From the set of RSV, only top r most similar documents were returned to user.

The following is an example of query expansion process. We enter the query term "diarrhea disease"[1] and after the query expansion we acquire a set of expanded terms: diarrhea disease, diarrhea, runs, diarrhoea. Then, we count the term of frequency of each expanded term in the document. After that, we calculate RSV by using the formula shown below.

$$\begin{aligned} &\text{Query} && = \text{diarrhea disease} \\ &\text{Expanded Terms} && = \{\text{diarrhea disease, diarrhea, runs, diarrhoea}\} \\ &\text{tf}_{diarrheadisease} && = 0 \\ &\text{tf}_{diarrhea} && = 6 \\ &\text{tf}_{runs} && = 1 \\ &\text{b}_{jk} && = \tfrac{tf(q_{ik}, d_j)}{1+tf(q_{ik}, d_j)} \\ &\text{q}_i && = \{1, 0.8, 0.64, 0.64\} \\ &\text{Vector } d_j && = \{\text{b}_{j1}, \text{b}_{j2}, \text{b}_{j3}, \text{b}_{j4}\} \\ & && = \{\tfrac{0}{1+0}, \tfrac{6}{1+6}, \tfrac{1}{1+1}, \tfrac{0}{1+0}\} \\ & && = \{0,0.86,0.5,0\} \\ &\text{RSV}_{doc} && = \{(0 \times 1) + (0.86 \times 0.8) + (0.5 \times 0.64) + (0 \times 0.64)\} \end{aligned}$$

---

[1] In this paper, to illustrate our examples, we use the English translation of Thai words actually used in the experiment, for convenience.

4.2.2 Browse Capabilities

Once the search is complete, browse capabilities provide the user with the capability to determine which items are of interest and select those to be displayed [6]. This is carried out in the two components of Browse Capabilities: Merging and Ranking components.

Merging

The retrieved URLs will be merged according to the Boolean operators the users specified and ranked based upon predicted relevance values. When users submit query Q which composed of $Term_1$ to $Term_i$ with operator AND | OR | NOT between them, we will merge the resulting expanded terms based on Boolean logic, as follows:

$$Q = Term_1 \oplus Term_2 \oplus Term_3 \oplus \cdots \oplus Term_i$$

where: $\oplus$ is the element of set {AND, OR, NOT}

For each term in user query, called $Term_i$, we apply the query expansion technique. As a result of expansion, we acquire the expanded list of each $Term_i$ as follows:

$$\begin{aligned}
\text{Expanded term of } Term_1 &= Term_{11}, Term_{12}, Term_{13}, \cdots, Term_{1j(1)} \\
\text{Expanded term of } Term_2 &= Term_{21}, Term_{22}, Term_{23}, \cdots, Term_{2j(2)} \\
\text{Expanded term of } Term_3 &= Term_{31}, Term_{32}, Term_{33}, \cdots, Term_{3j(3)} \\
&\vdots \\
\text{Expanded term of } Term_i &= Term_{i1}, Term_{i2}, Term_{i3}, \cdots, Term_{ij(i)}
\end{aligned}$$

where: i = number of query terms
j = number of expanded term of query term i
$Term_i$ = each term in user query Q
$Term_{ij}$ = the expanded term number j of $Term_i$

From the expanded list for each $Term_i$, the expanded term, say $Term_{ij}$ is selected and submitted to SANSARN search engine. The result of each query is R_{ij}. Consider $Term_i$ which has j expanded terms; we will have j sets of result from this step.

$$\begin{aligned}
\mathrm{R}_{i1} &= \mathrm{result}(Term_{i1}) \\
\mathrm{R}_{i2} &= \mathrm{result}(Term_{i2}) \\
&\vdots \\
\mathrm{R}_{ij} &= \mathrm{result}(Term_{ij})
\end{aligned}$$

then

$$\mathrm{R}_i = R_{11} \cup R_{12} \cup R_{13} \cup \cdots \cup R_{1i}$$

where: R_i = the set of relevant URL to $Term_i$

Finally, we apply the following formula to R_i depending on the operater between terms which is intersection, set union and set difference procedures for AND, OR and NOT, respectively, as shown in Figure 3.

$$result(URL) = R_1 \oplus R_2 \oplus R_3 \oplus \cdots \oplus R_i$$

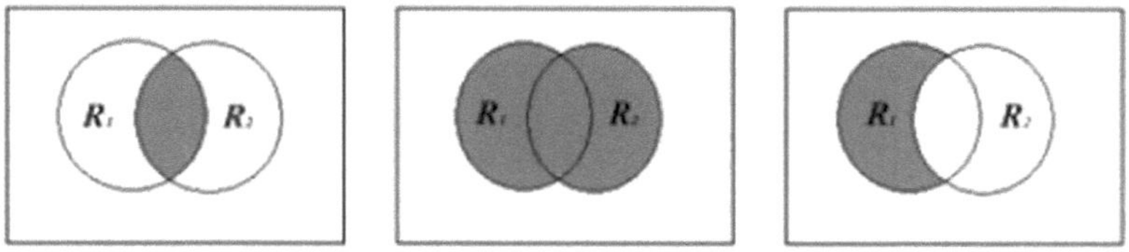

Fig. 4.3. The Boolean operators AND, OR and NOT are implemented using set intersection, set union and set difference procedures, respectively

Ranking

This step calculates and returns the n highest retrieval status values for a single query. With the ranking based upon predicted relevance values, the summary displays the relevance score associated with the item. The relevance score is an estimate of the search system on how closely the item satisfied the query statement. This reduces the user overheads by allowing the user to display the most likely relevant items first.

We use the following expression to compute new weight of each URL:

$$weight(URL_i) = \sum_{j=1}^{n} \sum_{k=1}^{n} weight(T_{jk})$$

where n is total number of user query term which is found in URL_i
m is total number of expanded term of query term j which is found in URL_i
$weight(T_{ik})$ is weight of expanded word k of user query term j

4.3 Results and Evaluation

4.3.1 Weight and Threshold Measurement

In the query expansion process, there are three important variables: W_{syno} or synonyms, W_{rel} or relate words and threshold. Based on our analysis and testing, two levels of expansion is the best expansion value. Because of these factors,

6 values of relationship are possible = {synonym, synonym ⇒ synonym, relevance, relevance ⇒ synonym, synonym ⇒ relevance, relevance ⇒ relevance}

Applying the weight calculation method to the set of relationships, we have:

Assume W_{syno} = x and W_{rel} = y

A set of weight results = x, x^2, y, yx, xy, y^2

We rank the threshold values for all elements in the above set, as shown below:

$$1 > x > x^2 > y > y_x, y_x > y^2 > 0$$

From the above formula, 1 is the value of original query. We tentatively assign 0.8 for x and 0.6 for y. The purpose of this step is to show how we can derive the best value of threshold. We consider 5 different values of threshold which can specify the degree of how relevance of the expanded words. If threshold is high, this means that we will expand only to the words that are similar to the original query. From the results set of each threshold value, the precision of a query can be calculated. This was carried out for all of the queries and all 5 different threshold values. Table 2 presents five selected values of threshold with their average precision values.

According to Table 2, we found that threshold=0.5 is the best threshold value. From this threshold, our system precision is 97%. While threshold=0.8 gives too less relevant documents and threshold < 0.36 drop the precision. When we incorporate our query expansion to SANSARN search engine, we define the difference in the average precision as shown in the following table.

We considered three different values of thresholds: 0.4, 0.5 and 0.6. The average precisions of SANSARN with query expansion is 96%, 97% and 93%, which differ +3.49%, +4.07% and +2.09% from the original precision of SANSARN system.

Table 4.2. Different values of thresholds and their performance output

Threshold	Avg. Precision(%)
$1 \geq$ threshold > 0.8	93
$0.8 \geq$ threshold > 0.64	93
$0.64 \geq$ threshold > 0.6	95
$0.6 \geq$ threshold > 0.48	97
$0.48 \geq$ threshold > 0.36	96
$0.36 \geq$ threshold > 0	Too many errors

Table 4.3. Percentage of differences from original SANSARN search engine acording to various values of thresholds

Threshold	Difference from SANSARN (%)	Remark
$1 \geq$ threshold > 0.8	0	SANSARN results
$0.8 \geq$ threshold > 0.64	0	Too less expansion
$0.64 \geq$ threshold > 0.6	+2.09	threshold = 0.62
$0.6 \geq$ threshold > 0.48	+4.07	threshold = 0.5
$0.48 \geq$ threshold > 0.36	+3.49	threshold = 0.4
$0.36 \geq$ threshold > 0	-	Too much expansion

4.3.2 Usefulness Measurement

Usefulness [7] is an effectiveness measure which compares two retrieval algorithms; given two algorithms, A and B, $u(A,B)$ expresses whether A is more effective than B or vice versa. A positive value indicates that B is more effective than A; a negative value signifies that B is less effective than A. The measurement is specifically suited to measure IR algorithm performance in a WAN environment. The notion of item relevance in this effectiveness measure does not have to be an absolute binary (relevant / non-relevant) judgment. Instead, users may rank information relevance relative to other items. The viability of usefulness was demonstrated in [8] and its advantages to traditional measures are evident in [9].

Usefulness uses all preference relations [10], [11], [12], contained in each user's specification of the relative relevance of the items delivered. In the classical preference relation $<_p$ [10], a person's preference of one message d' to d another is represented as: $d <_p d'$. This preference is assumed to be based on the relative relevance of the two messages to the person p. These judgments are made by a person after reviewing d and d'. Partial preference relations $\prec_p$ [12] are similar to the classical preference relations, i.e. $d \prec_p d'$ also indicates that person **p** prefers message d' to d.

In order to evaluate our query expansion system using the usefulness measure, we first need to derive the relative relevance assessments from the search results of original SANSARN (set A) and search results of SANSARN incorporated with our query expansion system (set B). We will consider only the first r documents retrieved in both set A and set B. From the results of $2r$ documents, we have the two users access the documents by marking the documents with the grades (5, 4, 3, 2, 1). The grades Grade(d,q) indicate the degree of relevance of each document d with respect to the query q. The higher the grade, the more relevant the document is. The scale for the marking is shown in Table 4.

In order to explain how we calculate the usefulness, what follows is the explanation and some examples from set A and set B. Table 5 is the result R_A(D,q,r) of original SANSARN and the result R_B(D,q,r) of SANSARN incorporated with query expansion method.

Table 4.4. Scale for marking retrieved documents

Meaning	Grade
Extremely relevant	5
Relevant	4
Marginally relevant	3
Irrelevant	2
Extremely irrelevant	1

Table 4.5. Examples of results R_A(D,q,r) and R_B(D,q,r)

query	Rank 1	Rank 2	Rank 3	Rank 4	Rank 5	Method
jaundice (R_A)	D1	D7	D10	D4	D2	original SANSARN
jaundice (R_B)	D3	D4	D5	D6	D7	SANSARN + QE
diarrhea (R_A)	D1	D2	D3	D4	D5	original SANSARN
diarrhea (R_B)	D8	D9	D11	D10	D12	SANSARN + QE
tuberculosis (R_A)	D7	D1	D2	D5	D9	original SANSARN
tuberculosis (R_B)	D3	D4	D5	D6	D7	SANSARN + QE
tumor (R_A)	D2	D7	D5	D11	D10	original SANSARN
tumor (R_B)	D1	D12	D3	D4	D13	SANSARN + QE

The answer sets R(D,q,r) of query are acquired by the following merge:

$$R(D, q, r) = R_A(D, q, r) \cup R_B(D, q, r)$$

The example results of the above merge are shown in Table 6.

Table 4.6. Example of R(D,q,r)

Query	R(D,q,r)
jaundice	{d3,d4,d6,d7,d8}
diarrhea	{d8,d10,d15,d16,d11}
tuberculosis	{d4,d1,d7,d2,d6}
tumor	{d1,d2,d3,d4,d5}

This set of documents $R(D, q, r)$ contains at most $2r$ documents. Since each algorithm return r of less than r documents, we set the relevance score for each document. Only documents that have relevance score of more than 2 were selected and ranked. If more than one document have the same relevance score, we specify the preference manually. Hence, the preferences $\prod_p^t$ derived from relevance information evaluated by the users are as follows:

$$\prod_{p}^{t} = \{dj > qk\ di \mid Grad(dj, qk) > Grad(di, qk)\} \tag{4.1}$$

Note that the preference (d, d') means that user judges d to be more useful than d'. The result is shown in Table 7.

Table 4.7. The Preferences (d, d')

Query	$R^2(D,q,r) \cap \prod p$
jaundice	{(d3,d4),(d3,d6),(d3,d7),(d3,d8),(d4,d6),(d4,d7),(d4,d8),(d6,d7),(d6,d8),(d7,d8)}
diarrhea	{(d8,d10),(d8,d15),(d8,d16),(d8,d11),(d10,d15),(d10,d16),(d10,d11),(d15,d16),(d15,d11),(d16,d11)}
tuberculosis	{(d4,d1),(d4,d7),(d4,d2),(d4,d6),(d1,d7),(d1,d2),(d1,d6),(d7,d2),(d7,d6),(d2,d6)}
tumor	{(d1,d2),(d1,d3),(d1,d4),(d1,d5),(d2,d3),(d2,d4),(d2,d5),(d3,d4),(d3,d5),(d4,d5)}

Table 4.8. The preferences $\prod_A$ of original SANSARN

Query	Preferences of original SANSARN ($\prod_A$)
jaundice	{(d1,d7),(d1,d10),(d1,d4),(d1,d2),(d7,d10),(d7,d4),(d7,d2),(d10,d4),(d10,d2),(d4,d2)}
diarrhea	{(d1,d2),(d1,d3),(d1,d4),(d1,d5),(d2,d3),(d2,d4),(d2,d5),(d3,d4),(d3,d5),(d4,d5)}
tuberculosis	{(d7,d1),(d7,d2),(d7,d5),(d7,d9),(d1,d2),(d1,d5),(d1,d9),(d2,d5),(d2,d9), (d5,d9)}
tumor	{(d2,d7),(d2,d5),(d2,d11),(d2,d10),(d7,d5),(d7,d11),(d7,d10),(d5,d11),(d5,d10),(d11,d10)}

In the same way, we can derive the preferences $\prod_A$ from original SANSARN system and the $\prod_B$ from SANSARN incorporated with our query expansion system. The preferences $\prod_A$ is already specified as the output ranking from SANSARN. The preferences $\prod_B$ is calculated from RSV value as shown below:

$$\prod_{p} = \{(d, d') \mid RSV(d) > RSV(d')\}$$

Table 8 and Table 9 show the preferences $\prod_A$ of original SANSARN and the preferences $\prod_B$ of SANSARN with our query expansion system respectively.

After we have set each preference, we have to calculate the satisfied preference. The satisfied preference $R^2(D,q,r) \cap \prod p \cap \prod x$ is the preference $\prod x$ that matches the answer set preference $R^2(D,q,r) \cap \prod p$. The satisfied preferences are shown in Table 10.

Table 4.9. The preferences $\prod_B$ of SANSARN with our query expansion system

Query	Preferences of SANSARN with our query expansion ($\prod_B$)
jaundice	{(d3,d4),(d3,d5),(d3,d6),(d3,d7),(d4,d5),(d4,d6),(d4,d7),(d5,d6),(d5,d7),(d6,d7)}
diarrhea	{(d8,d9),(d8,d11),(d8,d10),(d8,d12),(d9,d11),(d9,d10),(d9,d12),(d11,d10),(d11,d12),(d10,d12)}
tuberculosis	{(d3,d4),(d3,d5),(d3,d6),(d3,d7),(d4,d5),(d4,d6),(d4,d7),(d5,d6),(d5,d7),(d6,d7)}
tumor	{(d1,d12),(d1,d3),(d1,d4),(d1,d13),(d12,d3),(d12,d4),(d12,d13),(d3,d4),(d3,d13),(d4,d13)}

Table 4.10. Satisfied Preferences

Query	$R^2(D,q,r) \cap \prod p \cap \prod A$ (original SANSARN)	$R^2(D,q,r) \cap \prod p \cap \prod B$ (SANSARN with our query expansion)
jaundice	{}	{(d3,d4),(d3,d6),(d3,d7),(d4,d6),(d4,d7),(d6,d7)}
diarrhea	{}	{(d8,d10),(d8,d11),(d10,d11)}
tuberculosis	{(d7,d2),(d1,d2)}	{(d4,d6),(d4,d7)}
tumor	{(d2,d5)}	{(d1,d3),(d1,d4),(d3,d4)}

Table 4.11. Values x, y, and y_i - x_i

Query	x_i	y_i	y_i - x_i
jaundice	0.00	0.60	0.60
diarrhea	0.00	0.20	0.20
tuberculosis	0.20	0.20	0.0
tumor	0.10	0.30	0.20

Next, in the following formula, the x_i and y_i denoted the portion of preferences satisfied by our algorithm and SANSARN without our algorithm respectively, as shown in Table 11.

$$x_i = \frac{R^2(D,q,r) \cap \prod p \cap \prod A}{R^2(D,q,r) \cap \prod p}$$

$$y_i = \frac{R^2(D,q,r) \cap \prod p \cap \prod B}{R^2(D,q,r) \cap \prod p}$$

Given the pairs $\{(x_i, y_i) | 0 <= i < k\}$, the sum of the positive ranks w_+ is computed as follows:

1. Calculate the differences $y_i - x_i$ and discard the differences that are equal to zero. They do not contribute to the comparison of A and B.
2. Rank the absolute values $|y_i - x_i|$ of the differences $y_i - x_i$ in increasing order. If there are ties ($|y_i - x_i| = |y_j - x_j|$), each $|y_i - x_i|$ is assigned the average value of the ranks for which it is tied.
3. Restore the signs of the $|y_i - x_i|$ to the ranks, obtaining signed ranks.
4. Calculate w_+, the sum of those ranks that have positive signs.

The following table shows the results of the above algorithm:

Table 4.12. Ranks of $|y_i$ - x$_i|$

Query	$\|y_i$ - x$_i\|$	sign	rank
jaundice	0.6	+1	1
diarrhea	0.2	+1	2.5
tumor	0.2	+1	2.5

The usefulness $u_{(A,B)}$ is defined to be the normalized deviation of w$_+$ from μ, the expectation of w$_+$ that is obtained when x_i and y_i have the same distribution. The value k is reduced to the number of valid experiments (without ties) which will be called k_o.

$$u_{A,B} = \frac{\omega_+ - \mu}{\mu}$$

where

$$\mu = \frac{k_0(k_0 + 1)}{n}$$

Since $|y_2 - x_2|$ and $|y_3 - x_3|$ are tied, they are assigned an average rank of 2.5 = (2+3)/2. From the ranked differences we obtain the sum of positive ranks, the usefulness of A with respect to B, and the error probability. Note that one experiment (query q_1) was discarded. Hence, the number of experiments k_0 is equal to 3 rather than 4.

$$\begin{aligned} \mathrm{w}_+ &= 1{+}2.5{+}2.5, \\ \mu &= 3{\times}(3{+}1)/4 = 3 \\ \mathrm{u}_{(A,B)} &= (6\text{-}3)/3 = 1 \end{aligned}$$

All steps described above are the algorithm used to calculate usefulness. Tables 5 - 12 show the results of a sample test set of document. In the real experiments we use the first 20 documents from both systems and two users access and validate the result sets. The query is the same query set using to measure precision. All usefulness experiments are testing against the threshold 0.5, which gave the best result of precision. The final result is shown in Table 13.

Table 4.13. u(ss,qe) Usefulness value of SANSARN search engine when incorporated with our query expansion method

Thershold	u(ss,qe)	Pk
0.5	0.53	0.02

From Table 13, a positive value indicates that SANSARN search engine, incorporated with query expansion is more effective (i.e. useful) than SANSARN system without query expansion. As mentioned earlier, a positive value $u(A, B)$ signifies that B is more useful than A. Conversely, a negative value $u(A, B)$ signifies that B is less useful than A. In addition, an error probability P_k is given which expresses how much we can depend on the value $u(A, B)$. The corresponding error probability values P_k are very small indicating that this usefulness value is reliable.

4.4 Further Application of Query Expansion

Apart from using the query expansion technique to improve information search on the web, it can also be used for other intelligent systems such as systems that support education on the web, i.e. intelligent tutoring systems (ITSs) or web-based tutoring systems. A well organized method for retrieving specific information or comparing user (or learner)'s queries with the database is promising. In order to test the application capability of our expansion system, we tried and used it with the learner's search for education content on the web.

In an intelligent tutoring system that we independently develop [13], if the learner wants to search for more information on some issues or topics such as iteration, the system will search in the database for not only iteration but other concepts related to it and present the results to the learners using the query expansion method. Also, the merging and ranking are applied since learners of tutoring systems are normally interested in the top ranked documents. Hence, our tutoring system has been augmented with the query expansion method for search and browse capabilities.

4.5 Summary of Results

In this paper, we have augmented a search engine with query expansion technique. Furthermore, we also applied the technique to another intelligent system independently developed, i.e. a tutoring system to make it complete for learners using the system. As for our experimental results, we found that 0.5 is the best threshold value of query expansion to synonyms and related words. According to this threshold value, our system's precision is 97%. However, in our case, the evaluation in terms of recall cannot be determined since we do not know how many documents are contained in the back-end search engine that we use.

Therefore, we use the usefulness measurement [7] to evaluate whether the expanded terms are useful. Usefulness measurement u compares the effectiveness of two retrieval algorithms. As a result, we found that incorporating the query expansion method in to SANSARN, an existing bilingual (Thai and English) search engine, improves the retrieval quality.

References

1. Renals, S., Abberly, D.: The THISL SDR system at TREC-9. In: Proc. 9th Text Retrieval Conference (TREC-9), Gaithersburg, MD (2000)
2. Qiu, Y., Frei, H.P.: Concept Based Query Expansion. In: Proc. of the 16th Int. ACM SIGIR Conf. on R&D in Information Retrieval, Pittsburgh (1993)
3. Brisebois, M.: Query Expansion using WordNet (2003), `http://www.grafnetix.com/thesaurus/QueryExpansionIntro/QueryExpansionIntro.html`
4. Potipiti, T., Sornlertlamvanich, V., Charoenporn, T.: Towards Building a Corpus-based Dictionary for Non-wordboundary Languages. In: Proc. of LREC 2000: Second International Conference on Language Resources and Evaluation, Athens, Greece (2000)
5. Frei, H.P., Qiu, Y.: Effectiveness of weighted searching in an operational IR environmen. Information Retrieval: von der Modellierung zur Anwendung, 41–54 (1993)
6. Kowalski, G.: Information Retrieval Systems: Theory and Implementation. Kluwer Academic Publishers, Dordrecht (1998)
7. Frei, H.P., Schäuble, P.: Determining the Effectiveness of Retrieval Algorithms. Information Processing and Management 27(2&3), 153–164 (1991)
8. Wyle, M.F., Frei, H.P.: Retrieval Algorithm Effectiveness in a Wide Area Network Information Filter. In: Proc. 14th ACM SIGIR Conference on Research and Development in Information Retrieval, Chicago, IL, pp. 114–122 (1991)
9. Frei, H.P., Schäuble, P.: The Perils of Interpreting Recall and Precision Values. Informatik-Fachberichte Nr. 289, 1–10 (1991)
10. Bollmann, P., Wong, S.K.M.: Adaptive Linear Information Retrieval Models. In: Proc. 10th ACM SIGIR Conference on Research and Development in Information Retrieval, New Orleans, LA, pp. 157–163 (1987)
11. Wong, S.K.M., Yao, Y.Y., Bollmann, P.: Linear Structure in Information Retrieval. In: Proc. 11th ACM SIGIR Conference on Research and Development in Information Retrieval, Grenoble, France, pp. 219–232 (1988)
12. Schäuble, P.: Information Retrieval Based on Information Structures. Informatik-Dissertatioin. VDF-Verlag, Zurich (1989)
13. Facundes, N., Thepruangchai, P., Sirinaovakul, B.: Knowledge Acquisition and Representation for a Tutoring System. In: Auer, M., Ferreira, G. (eds.) Proceedings of International Conference on Interactive Computer Aided Blended Learning (ICBL). Kassel University Press, Florianopolis, Brazil (2007)

5

Comparing Clustering on Symbolic Data

Alzennyr da Silva[1], Yves Lechevallier[1], and Francisco de Carvalho[2]

[1] Project AxIS, INRIA-Rocquencourt
Domaine de Voluceau, B.P. 105
78153 Le Chesnay cedex, France
{Alzennyr.Da_Silva,Yves.Lechevallier}@inria.fr

[2] Centro de Informatica - CIn / UFPE
Av. Prof. Luiz Freire, s/n, CDU
50740-540 Recife, Brazil
fatc@cin.ufpe.br

Although various dissimilarity functions for symbolic data clustering are available in the literature, little attention has thus far been paid to making a comparison between such different distance measures. This paper presents a comparative study of some well known dissimilarity functions treating symbolic data. A version of the fuzzy c-means clustering algorithm is used to create groups of individuals characterized by symbolic variables of mixed types. The proposed approach provides a fuzzy partition and a prototype for each cluster by optimizing a criterion dependent on the dissimilarity function chosen. Experiments involving benchmark data sets are carried out in order to compare the accuracy of each function. To analyse the results, we apply an external criterion that compares different partitions of a same data set.

5.1 Introduction

Cluster analysis groups data objects only on the bases of information found in the data that describes the objects and their relationships. The goal is that the objects within a group should be similar (or related) to one another and different from (or unrelated to) the objects in other groups. The greater the similarity (or homogeneity) within a group and the greater the difference between groups, the better or more distinct the clustering is [17] [3].

Cluster analysis techniques can be divided into hierarchical and partitional methods [21] [11] [10].

Hierarchical methods yield a complete hierarchy, i.e., a nested sequence of partitions of the input data. Partitional methods seek to obtain a single partition of the input data into a fixed number of clusters. Often, these methods look for a partition that optimizes (usually locally) a criterion function. To improve cluster quality, the algorithm is run a number of times with different starting points and the best configuration obtained from all the runs is used as the output clustering.

N. Nedjah et al. (Eds.): Intelligent Text Categorization and Clustering, SCI 164, pp. 81–94.
springerlink.com

Partitional methods can be divided into hard clustering and fuzzy clustering. Hard clustering furnishes a hard partition where each object of the data set is assigned to one and only one cluster. Fuzzy clustering [1] generates a fuzzy partition that furnishes a degree of membership between 0 (absolutely doesn't belong) and 1 (absolutely belongs) of each pattern in a given cluster. The advantage of the fuzzy clustering approach over hard clustering is that the former is less prone to local minima because it makes soft decisions at each iteration due to the use of membership functions.

Dynamic clustering algorithms [8] are iterative two-step hard clustering algorithms. Each iteration involves constructing the clusters and identifying a suitable representation (e.g. means, axes, probability laws and groups of elements) of each cluster. It locally optimises an adequacy criterion between the clusters and their corresponding representations. The c-means algorithm, with the cluster representation updated after all the objects have been considered for relocation, is a particular case of dynamic clustering with the adequacy criterion equal to the variance and the cluster representation equal to the center of gravity.

This optimisation process starts from a set of representatives or an initial partition and interactively applies an "allocation" step (the exemplars are fixed) in order to assign the individuals to the classes according to their proximity to the exemplars. This is followed by a "representation" step (the partition is fixed) where the exemplars are updated according to the assignment of the individuals in the allocation step, until the algorithm converges, when the adequacy criterion reaches a stationary value. The dynamic cluster algorithm converges and the partitioning criterion decreases at each iteration if the class exemplars are properly defined at each representation step.

Objects to be clustered are usually represented as a vector of quantitative measurements. However, due to the recent advances in database technologies, it is now common to record multivalue and interval data. Symbolic data analysis is a relatively new field that provides a range of methods for analyzing complex datasets. The main contribution of the present work is to provide a comparative analysis and empirical evaluation of the most well-known dissimilarity functions for clustering symbolic data. Despite the importance of this kind of study, the issue has virtually not been addressed in the literature. In more detail, our proposition is to compare, in the framework of a fuzzy symbolic c-means algorithm, the efficiency of dissimilarities designed for symbolic data.

The remainder of this paper is organised as follows. Section 5.2 presents Symbolic Data Analysis (SDA). Section 5.3 decribes the clustering method used in our approach. Section 5.5 presents the experimental framework, the data sets analysed, the evaluation method used and the results. Section 5.6 provides an extension of our work to text classification and categorization. Section 5.7 gives the concluding remarks and some suggestions for future work.

5.2 Symbolic Data Analysis

Symbolic Data Analysis (SDA) is a domain in the area of knowledge discovery and data management related to multivariate analysis, pattern recognition

Table 5.1. Example of a classical data table

ID	Weight (kg)	#children	Sex	Nationality	Scholarity
1	54.8	0	M	Brazilian	High school
2	78.9	3	F	Brazilian	High school
3	110.5	2	M	Brazilian	Elementary school
⋮	⋮	⋮	⋮	⋮	⋮
N	80.0	1	F	American	High school

and artificial intelligence. It aims to provide suitable methods (e.g. clustering, factorial techniques and decision trees) for managing aggregated data described through multi-value variables, where there are sets of categories, intervals, or weight (probability) distributions in the cells of the data table [1]. In this so-called symbolic data table, the rows are the symbolic objects and the columns are the symbolic variables [4].

To illustrate the potential of SDA, let's describe a practical example. Let's consider a table containing only classic variables. The nature of these variables can be: (i) quantitative (discrete or continuous), for example, number of children and weight (in kg), respectively, and (ii) qualitative (dichotomic, nominal or ordinal), for example, sex, nationality and scholarity, respectively (see table 5.1). The nature of the variable descriptions in a classical data table allows each value to assume one category or numerical value. In contrast, in a symbolic data table, each observation is not restricted to a single value. The weight attribute, for example, can be registered as an interval [40.1 : 79.0] representing the weights of a family or a group of individuals, or even of an individual whose exact weight is not known. This type of variable is called *symbolic interval variable.* Symbolic data tables form the main input of the Symbolic Data Analysis. They are defined in the following manner: the columns of the input data table are variables that are used in order to describe a set of units known as "individuals". The rows are symbolic descriptions of these individuals because they are not vectors of single quantitative or categorical values.

Suppose now there is a variable named "automobile make" representing an individual possession. For a specific individual, we could have the set $Chevrolet, Fiat, Ford$ representing three different makes of automobile belonging to this person. This type of variable is known as *symbolic multivalued qualitative variable.* A third type of symbolic variable is known as *symbolic modal variable.* These are multi-state variables that have frequencies, probabilities, or weights related to their values. For example, if three people have diabetes and only one does not, the variable representing the tendency for diabetes among this group of people could be expressed by the distribution

[1] For more details about SDA, see `www.jsda.unina2.it`

Table 5.2. Example of a symbolic data table

ID	Weight (kg)	Automobile make	Tendency for diabetes
1	[54.8: 70.1]	{Ford, Fiat}	[3/4 diabetic, 1/4 non-diabetic]
2	[77.0: 80.0]	{Ford, GM}	[1/2 diabetic, 1/2 non-diabetic]
3	[89.8: 101.9]	{Fiat}	[2/5 diabetic, 3/5 non-diabetic]
⋮	⋮	⋮	⋮
m	[45.7: 90.9]	{GM, VW}	[1/2 diabetic, 1/2 non-diabetic]

$[\frac{3}{4}diabetic, \frac{1}{4}non-diabetic]$ (see table 5.2). Generally, this kind of variable can take form of, for example, a histogram, a frequency or a probability distribution.

SDA has also provided partitioning cluster algorithms for symbolic data. The authors of [7] used a transfer algorithm to partition a set of symbolic objects into clusters described by weight distribution vectors. The authors of [9] presented a fuzzy c-means algorithm to cluster symbolic data described by different types of symbolic variables. The authors of [22] introduced a dynamic cluster algorithm for symbolic data considering context-dependent proximity functions where the cluster prototypes are weight distribution vectors. The author of [12] presented an iterative relocation algorithm to partition a set of symbolic objects into classes and to minimize the sum of the description potentials of the classes. The authors of [5] proposed a dynamic cluster algorithm for interval data where the prototype is defined by the optimization of a criterion based on the Hausdorff distance.

5.3 Fuzzy Symbolic c-Means Algorithm

The fuzzy symbolic c-means algorithm [9] is a non-hierarchical clustering method whose aim is to furnish a fuzzy partition of a set of patterns in c clusters $\{P_1, \ldots, P_i, \ldots, P_c\}$ and its corresponding set of prototypes $\{g_1, \ldots, g_i, \ldots, g_c\}$ such that a criterion J, which measures the fit between the clusters and their prototypes, is locally minimized. The criterion J is defined as:

$$J = \sum_{k=1}^{n} \sum_{i=1}^{c} (w_{ik})^m d(x_k, g_i) \tag{5.1}$$

In this formula, $x_k = (x_k^1, \ldots, x_k^j, \ldots, x_k^p)$ is a feature vector describing the kth individual, $g_i = (g_i^1, \ldots, g_i^j, \ldots, g_i^p)$ is the feature vector describing the prototype of cluster P_i, w_{ik} is the membership degree of pattern k in cluster P_i, $m \in]1, +\infty[$ is a parameter which controls the fuzziness of membership of each pattern k and $d(x_k, g_i)$ is a distance function which measures the dissimilarity between a couple of feature vectors.

In the SDA domain, as we are dealing with complex data, the feature j can also represent a multi-value feature such as an interval or a weight distribution.

If the feature j is of an interval type, we must represent its lower and upper bounds. In this case, we have: $x_k^j = [x_k^{jl}, x_k^{ju}]$ and $g_i^j = [g_i^{jl}, g_i^{ju}]$.

The algorithm starts by setting an initial membership degree for each individual k in each cluster P_i and alternates a representation step and an allocation step until convergence when the criterion J reaches a stationary value representing a local minimum. In the representation step, the membership degree of each individual k on cluster P_i is fixed and the prototypes are updated according to the following expression:

$$g_i = \left[(g_i^1, e_i^1), \ldots, (g_i^j, e_i^j), \ldots, (g_i^p, e_i^p)\right], where:$$

$$e_i^j = \frac{\sum_{k=1}^{n} (w_{ik})^m * \theta}{\sum_{k=1}^{n} (w_{ik})^m} \text{ and } \theta = \left\{ \begin{matrix} 1, if x_k^j = g_i^j \\ 0, otherwise \end{matrix} \right\}$$

In the allocation step, the prototypes are fixed and the membership degree w_{ik} of each individual k in each cluster P_i is updated according to the following expression:

$$w_{ik} = \frac{1}{\sum_{h=1}^{c} \left\{ \frac{d(x_k, g_i)}{d(x_k, g_h)} \right\}^{\frac{2}{m-1}}} \tag{5.2}$$

5.3.1 Schema of the Algorithm

The fuzzy symbolic c-means algorithm receives as input parameters: c (number of clusters), m ($1 < m < \infty$), $\epsilon > 0$ (stopping threshold), N_i (maximum number of iterations), N_e (number of executions). Its execution is described in the following steps:

1. **Initialisation**
   Initialise $w_{ik}(k = 1, \ldots, n$ and $i = 1, \ldots, c)$ such that $w_{ik} > 0$ and $\sum_{i=1}^{c} w_{ik} = 1$. Set $t = 1$
2. **Representation**
   Compute the prototypes g_i of class $P_i(i = 1, \ldots, c)$ using equation (2)
3. **Allocation**
   Update the fuzzy membership degree w_{ik} of individual k $(k = 1, \ldots, n)$ belonging to cluster $P_i(i = 1, \ldots, c)$ using equation (3)
4. **Stopping criterion**
   If $|J^{t+1} - J^t| \leq \epsilon$ or $t > N_i$ then STOP
   else set $t = t + 1$ and GO TO step 2

As the initialisation of the cluster centers is random, the parameter N_e is used to indicate the number of times the algorithm must be repetead to avoid dependency on the initial points.

5.4 Dissimilarity Functions Analysed

In order to compare the efficiency of the dissimilarity function for symbolic data, we have chosen five representative dissimilarity functions proposed in the literature of SDA. These functions are described below. In its formulas, the function $\mu()$ computes either the length (if the feature j is of an interval type) or the number of elements included in a set (if the feature j is of a categorical type). D_j indicates the domain of feature j. The function $c()$ computes the complement of the values assumed by the feature j of an individual x in the domain D_j. Thus, $x^j \cup c(x^j) = D_j$. The join operator ($\oplus$) computes the union area and is defined according to equation 5.3.

$$x_k^j \oplus g_i^j = \left\{ \begin{array}{c} [min(x_k^{jl}, g_i^{jl}), max(x_k^{ju}, g_i^{ju})],\ if\ interval \\ x_k^j \cup g_i^j,\ if\ categorical \end{array} \right\} \tag{5.3}$$

The meet operator ($\otimes$) computes the overlapping area and is defined as $x_k^j \otimes g_i^j = x_k^j \cap g_i^j$ for categorical features. For its definition on interval features, see Figure 5.1.

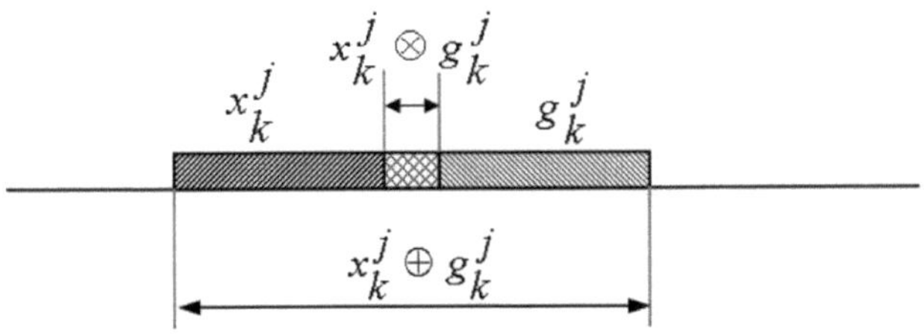

Fig. 5.1. Join and meet operators on interval features

5.4.1 Gowda and Diday's Dissimilarity Function

Gowda and Diday [13] proposed a dissimilarity measure for symbolic data that depends on three components, each of which deals with a specific dissimilarity aspect. This function and its components are defined as follows:

$$d(x_k, g_i) = \sum_{j=1}^{p} D_c(x_k^j, g_i^j) + D_s(x_k^j, g_i^j) + D_p(x_k^j, g_i^j), \text{ where:} \tag{5.4}$$

$$D_c(x_k^j, g_i^j) = \frac{\left|\mu\left(x_k^j\right) + \mu\left(g_i^j\right) - 2\mu\left(x_k^j \otimes g_i^j\right)\right|}{\mu\left(x_k^j \oplus g_i^j\right)} \text{ (content component)}$$

$$D_s(x_k^j, g_i^j) = \frac{\left|\mu\left(x_k^j\right) - \mu\left(g_i^j\right)\right|}{\mu\left(x_k^j \oplus g_i^j\right)} \text{ (span component)}$$

$$D_p(x_k^j, g_i^j) = \frac{\left|x_k^{jl} - g_i^{jl}\right|}{\mu\left(D_j\right)} \text{ (position component)}$$

5.4.2 Gowda and Ravi's Dissimilarity Function

Gowda and Ravi [14] proposed a dissimilarity measure for symbolic data based on Gowda and Diday's dissimilarity measure [13]. The main differences between these two measures are in the componentwise definition, as shown below:

$$d(x_k, g_i) = \sum_{j=1}^{p} D_c(x_k^j, g_i^j) + D_s(x_k^j, g_i^j) + D_p(x_k^j, g_i^j) \tag{5.5}$$

$$D_c(x_k^j, g_i^j) = cos\left[\frac{\mu\left(x_k^j \otimes g_i^j\right)}{\mu\left(x_k^j \oplus g_i^j\right)} * 90\right] \text{ (content component)}$$

$$D_s(x_k^j, g_i^j) = cos\left[\frac{\mu\left(x_k^j\right) + \mu\left(g_i^j\right)}{2\mu\left(x_k^j \oplus g_i^j\right)} * 90\right] \text{ (span component)}$$

$$D_p(x_k^j, g_i^j) = cos\left[\left(1 - \frac{\left|x_k^{jl} - g_i^{jl}\right|}{\mu\left(D_j\right)}\right) * 90\right] \text{ (position component)}$$

5.4.3 Ichino and Yaguchi's Dissimilarity Function

Ichino and Yaguchi [16] proposed a dissimilarity measure based on the join and meet operators (cf. equation 5.3 and figure 5.1). In this formula, the parameter γ is under condition $0 \leq \gamma \leq 0.5$

$$d_q(x_k, g_i) = \sqrt[q]{\sum_{j=1}^{p} \left[\psi(x_k^j, g_i^j)\right]^q}, \text{ where:} \tag{5.6}$$

$$\psi(x_k^j, g_i^j) = \frac{\phi(x_k^j, g_i^j)}{\mu(D_j)} \text{ and}$$

$$\phi(x_k^j, g_i^j) = \mu(x_k^j \oplus g_i^j) - \mu(x_k^j \otimes g_i^j) + \gamma(2\mu(x_k^j \otimes g_i^j) - \mu(x_k^j) - \mu(g_i^j))$$

This measure has a dependence on the chosen units of measurements. One solution to this problem is to remove this dependence by normalyzing the componentwise dissimilarity ψ. This solution was later proposed by De Carvalho [6] (cf. section 5.4.4).

5.4.4 De Carvalho's Extension of Ichino and Yaguchi's Dissimilarity Function

De Carvalho [6] extended the Ichino and Yaguchi dissimilarity measure (cf. section 5.4.3) by proposing a different componentwise dissimilarity measure defined as follows:

$$d_q^{'}(x_k, g_i) = \sqrt[q]{\sum_{j=1}^{p} \left[\frac{1}{p}\psi'(x_k^j, g_i^j)\right]^q} \tag{5.7}$$

$$\psi^{'}(x_k^j, g_i^j) = \frac{\phi(x_k^j, g_i^j)}{\mu(x_k^j \oplus g_i^j)}$$

Table 5.3. Agreement and disagreement indices

	Agreement	Disagreement	Total
Agreement	$\alpha = \mu(x_k^j \otimes g_i^j)$	$\beta = \mu(x_k^j \otimes c(g_i^j))$	$\mu(x_k^j)$
Disagreement	$\chi = \mu(c(x_k^j) \otimes g_i^j)$	$\delta = \mu(c(x_k^j) \otimes c(g_i^j))$	$\mu(c(x_k^j))$
Total	$\mu(g_i^j)$	$\mu(c(g_i^j))$	$\mu(D_j)$

Table 5.4. Microcomputer data set

ID	Microcomputer	Display	RAM	ROM	MP	Keys
0	Apple II	COLOR TV	48 K	10 K	6502	52
1	Atari 800	COLOR TV	48 K	10 K	6502	57 to 63
2	Commodore	COLOR TV	32 K	11 to 16K	6502A	64 to 73
3	Exidi sorcerer	B&W TV	48 K	4 K	Z80	57 to 63
4	Zenith H8	BUILT-IN	64 K	1 K	8080A	64 to 73
5	Zenith H89	BUILT-IN	64 K	8 K	Z80	64 to 73
6	HP-85	BUILT-IN	32 K	80 K	HP	92
7	Horizon	TERMINAL	64 K	8 K	Z80	57 to 63
8	Sc. Challenger	B&W TV	32 K	10 K	6502	53 to 56
9	Ohio Sc. II Series	B&W TV	48 K	10 K	6502C	53 to 56
10	TRS-80 I	B&W TV	48 K	12 K	Z80	53 to 56
11	TRS-80 III	BUILT-IN	48 K	14 K	Z80	64 to 73

In this formula, the function ϕ is defined as in Ichino and Yaguchi's dissimilarity measure. This function diverges only in the normalization, which is defined by applying the join operator (cf. equation 5.3 and figure 5.1) in the denominator of function ψ' definition. This measure is a metric with range between 0 and 1, and is not affected by a change of scale due to a linear transformation.

5.4.5 De Carvalho's Dissimilarity Function

De Carvalho [6] also proposed a new distance measure to symbolic data. His function is based on some agreement and disagreement indices (cf. table 5.3) and is defined as follows:

$$d_q(x_k, g_i) = \sqrt[q]{\sum_{j=1}^{p} \left[d(x_k^j, g_i^j)\right]^q}, \text{ where:} \tag{5.8}$$

$$d(x_k^j, g_i^j) = 1 - \frac{\alpha}{\alpha + \beta + \chi}$$

This measure is a straightforward extension of similarity measures for classical data matrices with binary variables.

Table 5.5. Fish data set

ID	Name	Length	···	Stomach/Muscle	Feeding
0	Ageneiosusbrevi	22.50 to 35.50	···	0.00 to 0.55	Carnivores
1	Cynodongibbus 1	19.00 to 32.00	···	0.20 to 1.24	Carnivores
2	Hopilasaimara 1	25.50 to 63.00	···	0.09 to 0.40	Carnivores
3	Potamotrygonhys	20.50 to 45.00	···	0.00 to 0.50	Carnivores
4	Leporinusfascia	18.80 to 25.00	···	0.12 to 0.17	Omnivores
5	Leporinusfreder	2.00 to 24.50	···	0.13 to 0.58	Omnivores
6	Dorasmicropoeus	19.20 to 31.00	···	0.00 to 0.79	Detritivores
7	Platydorascostas	13.70 to 25.00	···	0.00 to 0.61	Detritivores
8	Pseudoancistrus	13.00 to 20.50	···	0.49 to 1.36	Detritivores
9	Semaprochilodus	22.00 to 28.00	···	0.00 to 1.25	Detritivores
10	Acnodonoligacan	10.00 to 16.20	···	0.23 to 5.97	Herbivores
11	Myleusrubripini	12.30 to 18.00	···	0.31 to 4.33	Herbivores

Table 5.6. Wave data set

ID	Name	Position 1	···	Position 21	Group
0	wave_1_3	-2.99:2.63	···	-2.89:2.50	wave_1
1	wave_1_1	-2.09:2.83	···	-2.76:2.72	wave_1
2	wave_1_6	-1.70:2.89	···	-2.45:2.96	wave_1
3	wave_1_9	-2.36:2.69	···	-3.03:2.14	wave_1
4	wave_1_5	-2.27:2.33	···	-2.85:2.91	wave_1
⋮	⋮	⋮	⋮	⋮	⋮

5.5 Experimental Framework

5.5.1 Data Sets

To evaluate the dissimilarity measures mentioned above, our approach was carried out on four representative data sets. The microcomputer data set [18] (see table 5.4) is a classical benchmark in clustering literature. It consists of 12 individuals, two mono-value variables (*Display* and *MP*) and three interval[2] variables (*RAM, ROM* and *Keys*). The fish data set (see table 5.5) is a result of a study performed in French Guyana on fish contamination [2]. It consists of 12 individuals, 13 interval variables and one class variable that discriminates 4 types of animals: carnivores, detritivores, omnivores or herbivores. The wave data set (see table 5.6) is furnished in the SODAS[3] installation package and consists of 30 individuals described by 21 interval variables and 1 class variable that discriminates 3 types of waves. The zoo data set (see table 5.7) is available in the UCI Machine Learning Repository [20] and consists of 101 species of animals described by 15 Boolean, 1 numeric-value variable (number of legs) and 1 class variable that discriminates 7 types of animals.

[2] Single points are also treated as interval variables, known as degenerated intervals.

[3] SODAS stands for Symbolic Official Data Analysis System and can be downloaded at `http://www.ceremade.dauphine.fr/~touati/sodas-pagegarde.htm`

Table 5.7. Zoo data set

ID	Name	Hair	⋯	Catsize	Type
0	aardvark	1	⋯	1	1
1	antelope	1	⋯	1	1
2	bass	0	⋯	0	4
3	bear	1	⋯	1	1
4	boar	1	⋯	1	1
⋮	⋮	⋮	⋮	⋮	⋮

5.5.2 Evaluation Criterion

To compare the resulting partitions obtained by each dissimilarity function analysed, we adopted the an external criterion. This implies that we evaluate the results of a clustering algorithm based on how well the clusters match the gold standard classes.

In our experiments, we use the Corrected Rand index (CR) [15] as the external evaluation criterion. The CR index assesses the degree of agreement between an *a priori* partition and a partition furnished by the clustering algorithm. We used the CR index because it is not sensitive to the number of clusters in the partitions or to the distributions of the individuals in the clusters.

Let $U = \{u_1, \ldots, u_i, \ldots, u_R\}$ and $V = \{v_1, \ldots, v_j, \ldots, v_C\}$ be two partitions of the same data set having respectively R and C clusters. Their contingency table is depicted in figure 5.2. The CR is defined as:

$$\mathrm{CR} = \frac{\sum_{i=1}^{R}\sum_{j=1}^{C}\binom{n_{ij}}{2} - \binom{n}{2}^{-1}\sum_{i=1}^{R}\binom{n_{i.}}{2}\sum_{j=1}^{C}\binom{n_{.j}}{2}}{\frac{1}{2}\left[\sum_{i=1}^{R}\binom{n_{i.}}{2} + \sum_{j=1}^{C}\binom{n_{.j}}{2}\right] - \binom{n}{2}^{-1}\sum_{i=1}^{R}\binom{n_{i.}}{2}\sum_{j=1}^{C}\binom{n_{.j}}{2}} \tag{5.9}$$

	v_1	v_2	$\ldots$	v_C	
u_1	n_{11}	n_{12}	$\ldots$	n_{1C}	$n_{1.}$
u_2	n_{21}	n_{22}	$\ldots$	n_{2C}	$n_{2.}$
$\vdots$	$\vdots$	$\vdots$		$\vdots$	$\vdots$
u_R	n_{R1}	n_{R2}	$\ldots$	n_{RC}	$n_{R.}$
	$n_{.1}$	$n_{.2}$		$n_{.C}$	$n_{..} = n$

Fig. 5.2. Contingency table

where $\binom{n}{2} = \frac{n(n-1)}{2}$, n_{ij} represents the number of objects that are in clusters u_i and v_j, $n_{i.}$ indicates the number of objects in cluster u_i, $n_{.j}$ indicates the number of objects in cluster v_j and n is the total number of objects in the data set.

CR takes its values from the interval $[-1, 1]$, where the value 1 indicates perfect agreement between partitions, whereas values near 0 (or negatives) correspond to cluster agreement found by chance [19].

5.5.3 Results

For each dissimilarity function, the clustering method is run 50 times (until convergence to a stationary value of the adequacy criterion J) to avoid dependency on the initialisation phase. The best result, according to the corresponding adequacy criterion, is selected. Throughout these experiments, parameter m is set to 2, parameter γ is set to 0.5 and the order of power q is set to 2.

All the CR values obtained on each data set are given in table 5.8. In this table, the best values of the CR index are marked in bold. Values near to 1 represent good results. In other words, the results show the best solution reached by each dissimilarity function according to the CR values, i.e., these results are obtained from the partition of greatest similarity to the *a priori* partition. This partition is defined by the class attribute present in the input data set (and obviously not used in the clustering process).

Table 5.8. Corrected rand index values

Data set	Gowda and Diday's function	Gowda and Ravi's function	Ichino and Yaguchi's function	De Carvalho's function	De Carvalho's extension funcion
micro	0.909341	**1.000000**	**1.000000**	**1.000000**	**1.000000**
fish	**0.867470**	0.818681	**0.867470**	0.767278	0.767278
waves	**0.731042**	0.667145	**0.731042**	0.667145	0.667145
zoo	0.969127	0.966824	**0.972994**	**0.970124**	0.957612

To sum up, the following conclusions can be drawn from this empirical evaluation:

- For the case of data sets with only interval features (fish and wave data sets), the functions achieving the best results are those from Gowda and Diday and from Ichino and Yaguchi.
- For the set containing Boolean data (zoo data set), the best solutions were obtained by the dissimilarity functions from Ichino and Yaguchi and from De Carvalho. The latter is a straightforward extension of similarity measures for classical data matrices with binary variables, which gives this function some advantages in treating this kind of data.
- In the case of data sets with mixed features (microcomputer data set), all the analysed functions obtained good results. The only one that could not find the *a priori* partition was the dissimilarity function from Gowda and Diday.

- As we can notice, Ichino and Yaguchi's function outperfoms the others in most cases.

However, despite having obtained good general results, the functions analysed were only able to find the *a priori* partition for one data set (CR values equal to 1),i.e., place each individual in the correct cluster, which is known *a priori.* For the fish, wave and zoo data sets, the *a priori* class of each individual was obtained from the class variable presented in the data. For the microcomputers data set, this information was obtained from the literature [18].

5.6 Extension to Text Categorization and Text Classification

Nowadays, more and more information is circulating in a digital format in word processed documents. A vast area of application is related to the analysis of this kind of data, for example: classification of incoming e-mail according to predefined topics, grouping related documents based on their content without requiring predefined classes, or assigning documents to one or more user predefined categories.

Symbolic Data Analysis is a very powerful and extensible domain which can be applied in a wide range of disciplines, including information technology, health and social sciences. Primarily aimed at statisticians and data analysts, SDA is also appropriate for scientists working on problems involving large volumes of data such as structured or semi-structured text. SDA allows a more realistic description of the input units by taking into consideration their internal variation and their complex structure. SDA provides a better explanation of its results by an automatic interpretation which is closer to the users' natural language. SDA can thus be easily adapted to address text categorization and text classification issues.

Let's consider a large number of (semi-)structured documents in which texts are organized in blocks (for example, scientific articles made up of abstract, keywords, introduction, middle sections, conclusion, etc.). SDA provides a powerful tool to model this kind of data: a *symbolic data table* in which each line represents a document and each column is a *symbolic modal variable* (as described in section 5.2). Each variable represents a text block by means of frequencies of words from a controlled vocabulary containing p words (see table 5.9). For exam-

Table 5.9. Symbolic data table modelling (semi-)structured documents

id	abstract	keywords	...	conclusion
1	$(0.10)w_1, ..., (0.75)w_p$	$(0.41)w_1, ..., (0.06)w_p$	...	$(0.52)w_1, ..., (0.36)w_p$
2	$(0.16)w_1, ..., (0.15)w_p$	$(0.04)w_1, ..., (0.21)w_p$	...	$(0.00)w_1, ..., (0.02)w_p$
3	$(0.08)w_1, ..., (0.05)w_p$	$(0.01)w_1, ..., (0.30)w_p$	...	$(0.10)w_1, ..., (0.11)w_p$
$\vdots$	$\vdots$	$\vdots$	$\vdots$	$\vdots$
k	$(0.21)w_1, ..., (0.27)w_p$	$(0.19)w_1, ..., (0.77)w_p$	...	$(0.31)w_1, ..., (0.18)w_p$

ple, from the description of document 3 in table 5.9, we can say that the word w_1 represents 8% of the abstract content, whereas this same word represents only 1% of the keywords and 10% of the conclusion content.

To classify documents, a dissimilarity measure designed for modal symbolic objects can be applied to data formatted as in table 5.9. The clustering process will find a partition with clusters containing similar documents, i.e., documents sharing a set of common words. Moreover, if we are asked for a general description of each cluster, we can easily summarize the set of documents it contains by giving the description of the cluster prototype. The prototype of a cluster corresponds to its centre of gravity and is described by the mean value of word frequencies of documents belonging to the cluster.

5.7 Conclusion and Future Work

Many dissimilarity measures have been proposed in the literature on SDA, although an investigation comparing their applicability to real data has not attracted much attention. In this work we proposed a useful framework for comparing the efficiency of dissimilarity measures in the SDA domain. The conclusions are based on an external index for the analysis of clusters obtained by a fuzzy symbolic c-means algorithm with reference to four well-known benchmark data sets. The main challenge in our study stemmed from the need to implement and understand many different dissimilarity measures.

Our experiments show that more research is still needed in order to obtain a function that better approximates the *a priori* partition.

Future work may include: analyzing other dissimilarity functions and utilizing graphic tools to show how individuals are arranged into clusters at each step of the algorithm.

In order to enrich the initial findings of this study, more extensive experimentation involving artificial data sets and the analysis of each component of the functions has been planned. This work has already been started and will be the subject of future publications.

Acknowledgements

The authors are grateful to CAPES (Brazil) and the collaboration project between INRIA (France) and FACEPE (Brazil) for their support for this research.

References

1. Bezdek, J.C.: Pattern Recognition with Fuzzy Objective Function Algorithms. Advanced Applications in Pattern Recognition. Springer, Heidelberg (1981)
2. Bobou, A., Ribeyre, F.: Mercury in the food web: accumulation and transfer mechanisms. Metal Ions in Biological Systems, 289–319 (1998)
3. Bock, H.: Classification and clustering: Problems for the future. New Approaches in Classification and Data Analysis, 3–24 (1993)

4. Bock, H.-H., Diday, E.: Analysis of symbolic data: exploratory methods for extracting statistical information from complex data. In: Studies in classification, data analysis, and knowledge organization. Springer, Berlin (2000)
5. Chavent, M., Lechevallier, Y.: Dynamical clustering algorithm of interval data: Optimization of an adequacy criterion based on hausdorff distance. In: Classification, Clustering and Data Analysis, pp. 53–59 (2002)
6. De Carvalho, F.A.T.: Proximity coefficients between boolean symbolic objects. In: New Approaches in Classification and Data Analysis, pp. 387–394. Springer, Heidelberg (1994)
7. Diday, E., Brito, M.: Symbolic cluster analysis. Conceptual and Numerical Analysis of Data, 45–84 (1989)
8. Diday, E., Simon, J.: Clustering analysis. Digital Pattern Recogn., Commun. Cybern. 10, 47–94 (1976)
9. El-Sonbaty, Y., Ismail, M.A.: Fuzzy clustering for symbolic data. IEEE Transactions on Fuzzy Systems 6(2), 195–204 (1998)
10. Everitt, B.: Cluster Analysis. Halsted, New York (2001)
11. Gordon, A.D.: Monographs on Statistics and Applied Probability, Classification, 2nd edn., vol. 82. Chapman & Hall/CRC (1999) ISBN 1-58488-013-9
12. Gordon, A.D.: An iteractive relocation algorithm for classifying symbolic data. Data Analysis: Scientific Modeling and Practical Application, 17–23 (2000)
13. Gowda, K.C., Diday, E.: Symbolic clustering using a new dissimilarity measure. Pattern Recogn. 24(6), 567–578 (1991)
14. Gowda, K.C., Ravi, T.R.: Divisive clustering of symbolic objects using the concepts of both similarity and dissimilarity. Pattern Recognition Letters 28(8), 1277–1282 (1995)
15. Hubert, L., Arabie, P.: Comparing partitions. Journal of Classification 2, 193–218 (1985)
16. Ichino, H., Yaguchi, M.: Generalized minkowski metrics for mixed feature-type data analysis. IEEE Transactions on Systems, Man and Cybernetics 24(4), 698–708 (1994)
17. Jain, A.K., Murty, M.N., Flynn, P.J.: Data clustering: a review. ACM Computing Surveys 31(3), 264–323 (1999)
18. Michalski, R.S., Stepp, R.E.: Automated construction of classifications: Conceptual clustering versus numerical taxonomy. IEEE Transactions on Pattern Analysis and Machine Intelligence 5(4), 396–409 (1983)
19. Milligan, G.W.: Clustering validation: Results and implications for applied analysis. Clustering and classification, 341–375 (1996)
20. Newman, D.J., Hettich, S., Blake, C.L., Merz, C.J.: UCI repository of machine learning databases (1998), `http://www.ics.uci.edu/~mlearn/MLRepository.html`
21. Spaeth, H.: Cluster analysis algorithms. John Wiley and Sons, New York (1980)
22. Verde, R., De Carvalho, F., Lechevallier, Y.: A dynamical clustering algorithm for symbolic data. In: Tutorial on Symbolic Data Analisys, GfKl Conference, pp. 195–204 (2001)

6

Exploring a Genetic Algorithm for Hypertext Documents Clustering

Lando M. di Carlantonio[1,2] and Rosa Maria E.M. da Costa[1,3]

[1] Universidade do Estado do Rio de Janeiro – UERJ,
IME – Dept. de Informática e Ciência da Computação
Rua São Francisco Xavier 524, B–6°, CEP 20550–013
Rio de Janeiro – RJ – Brazil
[2] lando.mc@gmail.com
[3] rcosta@ime.uerj.br

Due to the increase in the number of WWW home pages, we are facing new challenges for information retrieval and indexing.

Some "intelligent" techniques, including neural networks, symbolic learning, and genetic algorithms have been used to group different classes of data in an efficient way.

This chapter describes a system for cluster analysis of hypertext documents based on genetic algorithms.

The system's effectiveness in getting groups with similar documents is evidenced by the experimental results.

6.1 Introduction

Currently, the diversity of available information in the Internet demands the development of new techniques to allow the user to find in a simple, fast and efficient way the information he desires.

Clustering is a widely used technique in data mining applications for discovering patterns in underlying data. Most traditional clustering algorithms are limited to handling datasets that contain either numeric or categorical attributes [1].

The cluster analysis to search documents is one of the interests of the data mining of texts, which objective is to group documents that have similar contents [2], [3], [4], [5]. In this case, the use of cluster analysis optimizes the classification of the information and provides groups with similar items, instead of a list of addresses, as offered by the available sites of search [6], or research works, [7], [8], [9]. In a clustering problem one has to partition a set of elements into homogeneous and well-separated subsets. The creation of such groups is a NP–complete problem.

Genetic algorithms (GAs) have been applied to this class of problems successfully; mainly in the cluster analysis of numerical data [10]. In this work, we describe a novel application of GA, which is based on the technique proposed by Hruschka&Ebecken [10] that developed an algorithm to group numerical data.

N. Nedjah et al. (Eds.): Intelligent Text Categorization and Clustering, SCI 164, pp. 95–117.
springerlink.com

From the basis of this algorithm, we developed and evaluated a system of cluster analysis of hypertexts documents called SAGH (Genetic Analytical System of Grouping Hypertexts). This system is specific to group html documents.

In the present chapter we present the main characteristics of the cluster analysis and the GAs techniques. In section 6.4, we describe the SAGH system. The section 6.5 presents the experimental results, followed by the conclusions and future works in section 6.6.

6.2 Cluster Analysis of Data

The objective of the cluster analysis applied to a data set is to create groups or clusters, based on the similarity among the components. This task defines the number of the groups and its elements, ensuring that the similar data remains together in a group.

The objective of cluster analysis is not to classify, since the cluster analysis considers that the number and shape of the groups are unknown a priori, whereas the classification considers that the groups are known [11].

6.2.1 Basic Concepts

Cluster analysis is defined as: from a set of n objects $X = X_1, X_2, \dots, X_n$, where each X_i that pertain to $\Re^p$ is a vector with p dimensions, which measures the attributes of the object. These must be clustered in a way which the groups $C = C_1, C_2, \dots, C_k$ are mutually disjoint, where k is a priori unknown value and represents the number of clusters [10]. In this case, the following conditions must be considered:

a) $C_1 \cup C_2 \cup \dots \cup C_k = X$;
b) $C_i \neq \emptyset, \forall\, i,\ 1 \leq i \leq k$;
c) $C_i \cap C_j = \emptyset, \forall\, i \neq j,\ 1 \leq i \leq k$ e $1 \leq j \leq k$.

Dificulties Finding the Ideal Clustering

Finding the ideal clustering of a data set is only computationally possible when the number of objects and the number of clusters are small because the problem is NP–complete. In this sense, the amount of distinct partitions where the data can be divided grows approximately as $k^n/k!$ [10].

As an example, for 25 objects grouped in 5 groups, there are a total of 2.436.684.974.110.751 possible different partitions. In addition, considering all the partitions involved from 1 to 5 groups, this number can go up to 4 x 10^{18} distinct partitions [12].

Data Structures

The data structures mostly used in cluster analysis are [13]:

a) data matrix: where there is a line for each object and a column for each considered attribute of the object;

b) dissimilarity matrix: n x n symmetric matrix where each element stores the distance between pairs of objects.

Types of Data

The types of data which cluster analysis deals with are [11]:

a) metric (quantitative): they represent measures, continuous or discrete, in a scale; they possess a unit (ex: s, m, kg, etc.);
b) non-metric (qualitative): they represent states of a finite and small set. They can be binary (ex: 0 or 1, yes or no), nominal ones (categorical) (ex: red, blue, yellow, etc.), or ordinal (ex: Sunday, Monday, Tuesday, etc.).

Measures of Similarity

The measures of similarity indicate the degree by what the objects have common characteristics.

In general, the majority of the measures of similarity explore the linear distance where the most similar objects are those that possess the smallest distance between them. Dissimilar objects are those with the greatest distance between them.

For the metric data, the measures of similarity are based on the following equation [13] (1):

$$d(i,j) = (|x_{i1} - x_{j1}|^q + |x_{i2} - x_{j2}|^q + \cdots + |x_{ip} - x_{jp}|^q)^{\frac{1}{q}} \tag{6.1}$$

Each parcel shows the difference between one attribute of the objects i e j. When q assumes the value of 1, it has the distance call to city–block or Manhattan. When it assumes the value of 2, we have the traditional and more frequent used Euclidean distance. For the other positive and integer values of q, the distance is known as Minkowski.

Classification of Methods

According to Han&Kamber [13], the methods of cluster analysis can be classified as:

a) partitioning method: this method creates an initial set of k sets, where k is the number of partitions desired; then, it moves objects of a cluster to another one through a technique of relocation, aiming to improve the partitioning;
b) hierarchical method: it creates a hierarchic decomposition for a set of objects. They can be agglomerative or divisive, in accordance with the form in which the decomposition is made;
c) density–based method: this method groups objects in function of the object density in the considered space;
d) grid–based method: it divides the object space in a finite number of cells that form a grid structure, and then executes a cluster analysis on this structure;

e) model–based method: this method generates a model for each one of the clusters and then it finds the distribution for the data that better is adjusted to the models created.

The Ideal Method

An ideal method of cluster analysis should have the following requirements [13]:

a) to find clusters with arbitrary shape, and of varied densities and sizes;
b) to deal with different types of data;
c) to reach the same results independently of the objects presentation order;
d) to handle objects with high number of dimensions;
e) to work with great sets of objects;
f) to generate interpretable, understandable and useful results;
g) to find the results even when there are unknown, missed or erroneous data;
h) to demand minimum knowledge to determine the input parameters;
i) to allow the definition of constraints to be observed in the cluster formation.

6.2.2 Cluster Analysis of Documents

The idea behind cluster analysis of documents is the same idea applied to numerical data: from a set of documents (objects) clusters are formed, where the number of clusters is not known a priori, and the documents of each cluster are more similar between themselves than in relation to documents from other groups.

What differentiates the two applications is the emphasis given to the step of preparation of the data, where equivalent structured representations for the documents are created. This representation is, normally, a vector containing the key words and a relevant measure of them in the document [14].

Relevance

Cluster analysis has been mainly studied for many years in statistics. With the development of data mining area, efforts have been made to develop methods capable of dealing efficiently with large datasets [13].

In order to extend the application of the methods created in data mining, many studies have been made to develop specific techniques that can be used to documents.

As a result of the globalizing use of electronic documents, beyond the increasing availability of information through the Internet, the cluster analysis of documents has acquired each day, more and more relevance [14].

Context

Cluster analysis of documents is one of the tasks of interest in text mining. Text mining deals with the discovery of new and potentially useful knowledge in text databases [14].

Steps of the Text Mining Process

The text mining process can be divided in the following steps [14]:

a) data preparation: it involves the sub-steps of information recovery and analysis of data;
b) execution of the desired text mining task: in this work, the task is exactly the cluster analysis, but other examples are: indexation, summarizing, categorization, etc.;
c) post-processing: it involves the evaluation of the discoveries and the validation of the results.

Data Preparation

The challenges faced in dealing with the text files to be surpassed for data preparation are [14]:

a) in general, texts do not possess structure: the majority of the traditional techniques of data mining need structured data to be applied;
b) commonly, texts possess thousands of words or terms: in order for the techniques of text mining to be executed with efficiency, it is necessary to reduce the number of terms by eliminating irrelevant words and aggregating repeated terms or with equivalent meaning.

Information Recovery. The sub-step of information recovery is responsible for the attainment of documents, that is, amongst existing documents in the available repository (in the case of this work, the Internet), those which are relevant to the criteria defined by the user are selected. The foundation of the utilized test is based on documents retrieve from the same service offered by search sites, such as Google [14].

Data Analysis. The data analysis sub-step converts the documents obtained in preceding phase into an appropriate representation for the accomplishment of the desired text mining task. In the case of this work: the cluster analysis [14]. In the data preparation, the most important sub-step is the data analysis. The information recovery was carried through by the search site and ended with the attainment of the necessary documents for the tests.

The procedures of data analysis used in this work are:

a) elimination of scripts and codes of language html: for the cluster analysis of documents, only the text of the document is relevant, being disqualified any text formatting codification, links specification, etc.;
b) elimination of numbers, symbols and punctuation: in general, this items have little or no importance for the text mining task;
c) conversion of accented characters in non-accented characters: the withdrawal of accents allows variants of certain words that possess accents to be added to the variant without the accent during the suffixes elimination.

d) conversion of the characters for one exactly standard (upper or lower case): the case folding procedure accelerates the comparisons carried through in the data analysis procedures;
e) elimination of words of one or two characters: this types of words, in rule, represent stop words and must be eliminated because they do not possess a discriminatory character. This procedure is made for performance profit because it is faster to evaluate the length of the word then to discover if it is present or not in a list of terms;
f) elimination of stop words: words without relevance such as prepositions, pronouns, articles, etc., are eliminated because they do not constitute knowledge in the texts. For the cluster analysis of documents, words which are present in diverse documents to be processed also are considered stop words because they do not contribute for the separation of documents;
g) elimination of suffixes (stemming): this procedure eliminates the possible existing suffixes in the words of the document. This allows reducing the search space dimensions;
h) accounting for the frequency of each term in each document and in the universe of documents to be clustered: there is a lot of repetition of terms inside each document. The frequency of one term gives the level of importance of this term in the subject of the document and in the set of documents. Through this procedure vectors–doc are obtained;
i) arrangement of vectors–doc: based on some criterion chosen by the user, vectors–doc are arranged;
j) determination of the terms that will form the p–dimensional search space: a number of terms, amongst the most frequent in the considered ordered vector–doc, is chosen for each ordered vector–doc;
k) construction of the matrix for cluster analysis: based on the dimension obtained, the corresponding frequencies of dimension terms are recovered inside each ordered vector–doc, creating the cluster analysis matrix lines;
l) normalization of the matrix: the matrix obtained in the previous procedure suffers normalization to compensate for the possible size difference between documents. This matrix is supplied as input for the cluster analysis task, which will find the number of clusters and the content of each cluster.

Execution of the Desired Task

After the accomplishment of the data preparation phase, the cluster analysis of documents task will be executed. The GAs technique, used to execute this step, will be detailed in section 6.3, while the method and the implementation details will be seen in section 6.4.

Main Methods for Cluster Analysis of Documents

The main methods of cluster analysis of documents are the Hierarchical Methods, the k–means method and the Self–Organizing Maps Method (SOMs) [14]. These methods are detailed in the next sections.

Hierarchical Methods

The hierarchical methods supply as result a structure called dendrogram, which is a diagram in tree format that describes the hierarchy of clusters [14].

In the root node of the tree, there is a partition that includes all vectors–doc submitted to the process; whereas in leaf nodes, the number of clusters is equal to the number of documents. The dendrogram can be created in two ways [13]:

a) agglomerative strategy: a cluster for each existing vector–doc is created. Then, repeatedly, clusters are joined until only one cluster containing all vectors–doc is obtained;
b) divisive strategy: this strategy is the inverse of the previous one, that is, a cluster that contains all vectors–doc is subdivided in two and thus successively divided until to have a cluster for each existing vector–doc.

k–means Method

The k–means method is a partitioning method. After the specification of a fixed number of clusters by the user, that is the k, the method chooses randomly k vectors–doc to be the initial centers of them. The others vectors–doc are distributed between the clusters in function of the existing similarity between them and the centers of the clusters.

If there is more than one element in the cluster, the center starts to be the centroid of the cluster, which is given by the average of vectors–doc present in the cluster. Having been made the distribution of all vectors–doc, new attributions are attemped to improve the objective function. The execution ends when there are no more alterations in the composition of each cluster [13].

Self–Organizing Maps Method

Self–organizing maps (SOMs) are artificial neural networks created by Kohonen according to connectionist paradigm of Artificial Intelligence [14].

In SOMs, the neurons that form the neural networks compete for the current vector–doc. Each neuron possesses a weight vector which indicates how the neuron is connected to the others [11]. During the processing, the neuron that possesses the most similar weight vector to the presented vector–doc is activated and has its weight readjusted in order to become more similar to the vector–doc current. The neighbor neurons to the winner neuron also suffer readjustment of their weights.

At the end of the process, it is expected that the association created between neurons and vectors–doc discloses the existing topology or arrangement between documents, forming a map [13]. This method is classified as a method based in models.

In the next section, the elementary concepts on GAs are examined.

6.3 Genetic Algorithms

Genetic Algorithms (GAs) are techniques of search and optimization inspired by genetics and the evolution of the species. They were presented by John Roland in 1975 and were created according to principle of the natural selection and survival of the fittest, established by Charles Darwin in 1859 in the book "On the Origin of Species" [15].

6.3.1 Basic Terminology

The terminology of the GAs is based on terms used in biology, whose adopted meanings are [15]:

a) population: set of potential solutions for the problem;
b) generation: each step of the evolution process;
c) chromosome: structure of data that stores a solution for the problem in a codified form;
d) individual: each member of the population. It is represented by the chromosome and its respective fitness, that, many times, is given by the objective function to be maximized;
e) gene: each one of the parameters codified in the chromosome;
f) allele: each value that one determined gene can assume;
g) phenotype: solution gotten by the decoding of the chromosome;
h) genotype: solution codified stored in the chromosome.

6.3.2 Conceptual Algorithm

The Conceptual GA is described for the following steps [15]:

1. Create initial population (Generation 0);
2. Evaluate initial population;
3. While stop criterion is not satisfied, do
   - Select individuals for the next generation;
   - Apply genetic operators in individuals selected;
   - Evaluate new population obtained;

6.3.3 Genetic Operators

The genetic operators used by the GAs are [15]:

Selection

This operator selects the best individuals, of bigger fitness, that exist in the current generation. These will give origin to the next population through the application of the crossover and mutation operators on them [15]. This operator searches the best points of the search space.

Crossover

The crossover operator is applied to the pairs of individuals selected by the selection operator [15]. It exchanges a randomly chosen part of the genetic material of the pair of individuals (parents A and B), supplying two new individuals (children C and D).

This operator allows the exploration of unknown regions of the search space. Its application does not guarantee that the new individuals will possess a better fitness than the original individuals.

Mutation

The mutation operator is applied in some individuals selected by the selection operator or in some of new individuals generated by the crossover operator [15]. It modifies, randomly, genes of the chromosome of these individuals. Its use improves the diversity of the population, besides being able to recuperate lost genetic material by the application of another genetic operator.

6.3.4 Stochastic Character

The GAs stochastic character allows the algorithm to surpass the existing minimums and/or local maximums in the search space, finding the minimum or global maximum of the problem [11].

6.3.5 Application Field

GAs are indicated for complex problems of optimization that do not possess efficient techniques of resolution, whose space of search presents many points of minimum and/or maximum [15].

6.3.6 Advantages

The GAs possess the following advantages in relation to others techniques [15]:

a) they work with continuous, discrete data and with combinations of them;
b) they make parallel searches in diverse regions of the search space;
c) they do not need derivatives or other auxiliary knowledge because they are guided by cost information or rewards;
d) it is not necessary to have a deep mathematical knowledge of the problem to be resolved;
e) they can optimize diverse parameters at the same time;
f) they are able to carry through optimizations even in search spaces of complex or complicated surface, with a reduced possibility to be imprisoned in local minimums;
g) they can be adapted to parallel computers;

h) they handle a codification of the set of parameters to be optimized and not of the parameters in question;
i) they generate as a result a list of excellent parameters instead of only one solution;
h) they utilize experimentally generated data and they tolerate incomplete data and noisy data;
j) they are implemented easily in computers;
k) they are easily reused in other problems because they are modular and portable, since the representation of the problem is separate of the evolution module;
l) they can be used with arbitrary constraints, optimizing various conflicting objective functions;
m) they can be easily integrated to other techniques and heuristics.

After the description of the GAs theoretical characteristics, the next sections present a system that explores the GAs to cluster hypertext documents.

6.4 The SAGH System

The SAGH (*Genetic Analytical System of Grouping Hypertexts*) is composed of seven modules. The first five modules manipulate and transform the documents aiming to construct a normalized matrix that will be used in the next steps. The sixth module carries through the cluster analysis; and the seventh, presents the results. This division in modules allows us to carry through analyses varying parameters without remaking all the steps.

The basic scheme of the SAGH System can be seen in the Fig. 6.1. Each module will be detailed in the next sections.

6.4.1 Creation of Vectors of Terms

This module examines the words that compose the text of each document, discarding items which define formatting, links, tables, image anchors, etc, from the html code.

Next, the module examines all words of the text and eliminates suffixes, according to Porter's algorithm [16] for the English language. Generally, as words that differ only in their suffixes have similar meanings, the elimination of them prevents the underestimation of radical keys. The module also offers the option to discard stop words or radical emptiness (stop words without suffixes). These three procedures reduce the number of significant words, increasing the performance of the system.

Each analyzed document generates a vector of terms (vector–doc) containing the radicals of the non-stop words and its respective number of occurrences in the text.

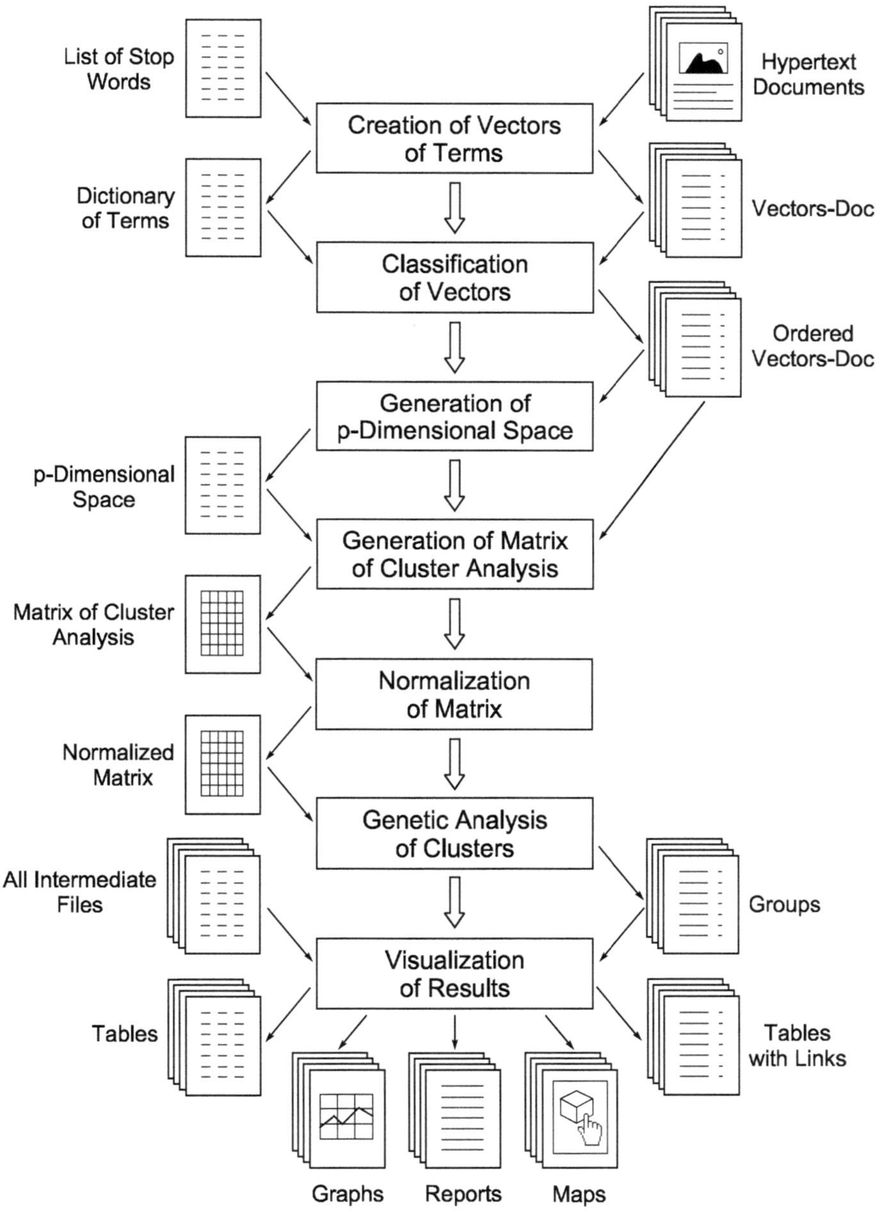

Fig. 6.1. The SAGH System

6.4.2 Classification of Vectors

The classification of vectors module sorts the vectors–doc generated in the previous module, from the following criteria:

a) the *tf* (*term frequency*): it is the number of repetitions of the term;
b) the *idf* (*inverse document frequency*): it is defined by the expression $idf(term) = \log(n/df_{term})$, where n is the number of documents to be grouped and, df_{term} is the number of documents where the term is present; and finally,
c) the most important criterion is the *tf*idf*, which is composed by the number of occurrences of the term in the document (*tf*) and a factor that decreases its relevance when it is present in a great number of documents or increase it otherwise (*idf*) [14]. If a word appears in almost all documents, it does not contribute to separate them.

This module sorts the vector–doc of each document, placing the most frequent word in the top of the list, creating the ordered vectors–doc. The *idf* and *tf*idf* criteria decrease the linear dependence between ordered vectors–doc when these have terms in common.

6.4.3 Generation of p–Dimensional Space

This module selects the more frequent i terms of each ordered vector–doc, creating the p–dimensional space, which will be considered to group the documents. The user specifies the value of i and the default value is 1.

We need a p–dimensional space with sufficient coordinates to avoid a great linear dependence and also, to not generate linear independent vectors that will be completely dissimilar.

In the SAGH system, the p–dimensional space is reached combining the ordered terms in each vector–doc from the criterion *tf*idf*, followed by the recovery of the term most frequent of each ordered vector–doc. With this, we will have only one vector (p–dimensional space) that will be generated by the terms that are in the top of each ordered vector–doc, eliminating the repetitions.

6.4.4 Generation of Matrix of Cluster Analysis

Based on the ordered vectors–doc and the p–dimensional space, this module constructs the cluster analysis matrix recovering the occurrence of each term of the p–dimensional space in each ordered vector–doc.

Each line of the matrix is a p–dimensional representation of a document, thus the columns are generated from the terms that compose p–dimensional space and the values associates to their occurrences inside the ordered vectors–doc. (See data matrix in subsection 6.2.1).

6.4.5 Normalization of Matrix

This module normalizes the values of the matrix, where each value of coordinate a_{ij} will be in the range [0..1]. The values of the coordinates are modified according to (2), where p is the number of columns in the matrix:

$$norm\_a_{ij} = \frac{a_{ij}}{\sqrt{\sum_{t=1}^{p} a_{it}^2}} \tag{6.2}$$

6.4.6 Genetic Analysis of Clusters

After the creation of the normalized matrix, the genetic analyzer of clusters module creates the groups, defining: the number of groups and their composition; its centroids; the documents distance from the centroids of its respective groups; and the main terms of each group.

Moreover, it calculates the distances of matrix and informs: the evolution of the fitness function through the generations, the online and offline performances of the GAs by generation [17], the best number of groups found in each generation and, optionally, each one of the populations evaluated during the system execution.

This module, specific for the analysis of document clustering, was developed based on the Hruschka&Ebecken [10] proposal, which described an algorithm for analysis of numerical data clustering.

The details of this method are described in the following sections.

Main Characteristics of the Hruschka&Ebecken Proposal

The main characteristics of this algorithm are:

- partitioning method based on GAs;
- chromosomes of constant size throughout the execution;
- fitness function based on the average silhouette width;
- calculation of dissimilarities through the Euclidean metric;
- stopped criterion based on generation number;
- selection from the roulette–wheel selection;
- elitism use;
- crossover and mutation operators oriented to group;
- random initial population;
- it does not need any input parameter;
- and it finds not only the number of groups in which the objects must be distributed, but also the composition of each one of them.

Chromosome Representation

The chromosome utilized is a one–dimensional array of size $n+1$, where n is the number of objects to be grouped. For each object i, between 1 and n, *Cromossomo[i]* identifies to which group (integer numbers) the object i belongs, while *Cromossomo[n+1]* stores the distinct number of groups that exist in the chromosome.

Therefore, the n first positions of the chromosome define the partitioning that it represents, whereas the position $n+1$ informs the existing number of groups in this partitioning.

Thus, considering k as the maximum number of groups where the objects can be distributed, the labels contained in *Cromossomo[i]* vary from 1 to k, and by the definition of cluster analysis seen in the subsection 6.2.1, k has to be lesser or equal the n.

An example can be seen in Fig. 6.2.

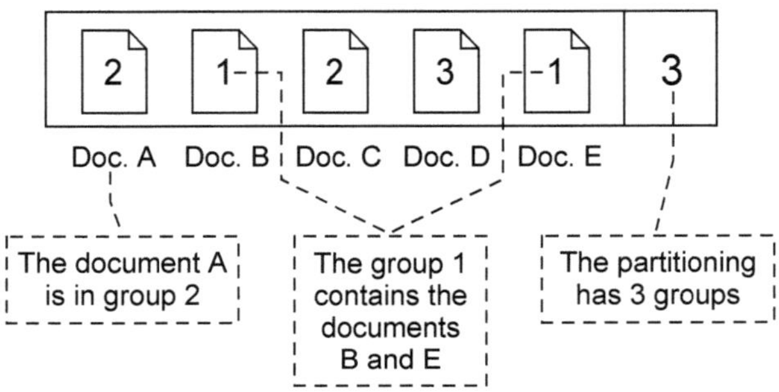

Fig. 6.2. A chromosome: partitioning of the documents + number of distinct groups

In a set of objects formed by 5 (n) elements, this chromosome represent a possible chromosome clustering the objects in 3 groups, where, for the purposes of this work, the objects are considerate as 5 documents, A, B, C, D and E. The chromosome indicates that group 1 possesses the documents B and E, group 2 possesses documents A and C and group 3 only possess document D. The information of the existing distinct number of groups in the chromosome, which are 3, is found in the position 6 ($n+1$) of the chromosome.

Objective Function

The objective function utilized is calculated initially by (3):

$$OF(chromosome) = \sum_{i=1}^{n} \frac{s(i)}{n} \tag{6.3}$$

Where n is the number of objects to be clustered and $s(i)$ is the average silhouette width, which for one determined object i, it is given by (4):

$$s(i) = \frac{b(i) - a(i)}{max\{a(i), b(i)\}} \tag{6.4}$$

Where $a(i)$ is the arithmetic mean of the distances of the object i to each one of the other existing objects in the group where object i is found.

In addition, *b(i)*, which represents the distance of object i to the next neighboring cluster, is given by: $b(i) = min\ d(i,\ C),\ C \neq A$, where $min\ d(i,\ C)$ is the smaller value of the arithmetic mean of the object i distances (pertaining to a cluster A) to all the pertaining objects to one determined cluster C. If the cluster only contains one element, it is chosen that $s(i)$=0. The value of $s(i)$ varies from -1 to 1.

After the calculation of the objective function for all the chromosomes of the population through the eq. 2, a linear normalization is applied to the values. Next, the values are organized in a descending order – in the original article the values are organized in an ascending order. Then, they are substituted by (5):

$$OF'(i) = \frac{[PopSize - R(i)] * [OF(max) - OF(min)]}{PopSize} \qquad (6.5)$$

Where $R(i)$ is the position of chromosome i in the order; $OF(max)$ and $OF(min)$ are, respectively, the greater and the minor values found for the objective function; and *PopSize* is the number of chromosomes in the population.

Selection

The individuals are selected by roulette–wheel selection.

The roulette–wheel selection can be made by the following manner:

a) from the population ranked in ascending order, it calculates the accumulated fitness of each one of the individuals;
b) it generates a random number r in the interval, [0, TotalSum], where TotalSum is the value of the addition of all the fitness of the existing individuals in the population;
c) it searches for a descending order of the objective function, selects the first individual with the value of superior accumulated fitness to the value of r;
d) it repeats the previous steps until obtains the number of individuals desired.

The roulette–wheel selection does not allow negative values for the objective function. Therefore, a constant number equal to one is added to each value of the objective function, modifying the interval of the average silhouette to $0 \leq s(i) \leq 2$. This is only made for the roulette–wheel selection, preserving the original values of the objective function.

The method also utilizes the elitist strategy of copying the best individual of each generation to the following generation. It guarantees that the best individual found until then will be present in the next generation, because it could have been lost after the application of the genetic operators.

Crossover

The chromosomes children are obtained by the following way:

a) two chromosomes are selected to be the parents A and B;
b) randomly, i groups of the father A are chosen to be copied for the son C, having as its limit the existing distinct number of groups in the father A;

c) the genes of father B whose respective genes of the father A did not have its labels chosen in item (b) are copied for son C;
d) the genes of son C that were not labeled by item (b) and (c) are placed in the closer (whose centroid is closer to the object) groups (existing in son C). It is observed that the centroids are not recalculated to each allocation.

For attainment of son D, it works in a analogical form, but considering A in the place of B and B in the place of A, and considering groups chosen by chance – item (b) – previously gotten in item (c).

Mutation

There are two genetic operators of mutation:

a) mutation for elimination of a group chosen by chance;
b) mutation to divide a group chosen by chance in two.

Eliminating Groups

Randomly, an existing group in the chromosome is chosen and eliminated, placing the objects contained in the group in the remaining groups in function of the proximity of the objects in relation to the groups centroids.

This operator is only applied if the chromosome represents more than two groups, preventing a chromosome where all the objects are in a same group, that is, without any objects division.

Dividing Groups

A group amongst the existing ones in the chromosome is chosen randomly and divided in two new ones (even though one of them remains with the same label, it is said two new ones due to the suffered alteration in the elements that compose the groups).

It becomes two references: the centroid of the chosen group and the object that is more distant of the group centroid. Then, the objects of the group are distributed in function of the smaller existing distance between each one and these two references.

Steps

The steps of the algorithm are:

a) generate an initial population composed of k-1 chromosomes, where k is the maximum number of groups in which the objects can be distributed, in the following manner:

- the first chromosome must possess 2 groups, the second 3 groups and successively, in a way which the *kth*-1 chromosome possesses k groups;
- given the number of groups that the chromosome must possess, distribute, randomly, all the objects between the groups that the chromosome represents;

b) evaluate the population;
c) apply linear normalization to values obtained for the objective function;
d) select individuals for proportional selection (roulette–wheel selection);
e) apply crossover in the first 50%, mutation 1 (elimination of a group) in the following 25% and mutation 2 (division of a group in two) in the remaining 25%, obtaining the new population;
f) if the stop criteria is satisfied, stop; otherwise, return to step (b).

Evaluation Method

The method proposed by Hruschka&Ebecken considers some requisites when it is compared to the ideal cluster analysis method described in the sub-section 6.2.1 [18]:

a) the method is more appropriate to find groups in the hyper-spherical shape. In a certain limit, it can find groups of diverse sizes;
b) the method only deals with metric data;
c) the method is capable of reaching the same results even with change in the order of presentation of objects;
d) the method performance gradually decreases with the increase of the dimension numbers of each object;
e) in the same way as mentioned before, the performance decreases with the increase of the number of objects;
f) the method offers a simple and easy to understand result;
g) the method possesses erroneous results when there are objects with wrong data and it does not work with absent values;
h) the method does not need any input parameter;
i) the method does not allow constraint definitions.

Challenges to Create the SAGH System

The faced challenges to create the SAGH system from the Hruschka&Ebecken algorithm involved:

- the determination of the search space;
- the creation of a mechanism to compensate the stochastic character of GAs (super population);
- the identification of a metric adjusted to the calculation of dissimilarities between documents and that is usable in the calculus of average silhouette width (cosine metric);
- the development of document transformation procedures in the normalized cluster analysis matrix;

- the choice of efficient data structures to manipulate the intermediate files and the generation of reports to justify the formed groups.

The used mechanism for the super population control runs the system repeatedly, saving the best individual of each execution to form the initial population of the final execution.

The calculation of the differences between two documents from the cosine metric is given by $\cos(d_1,d_2) = (d_1 \bullet d_2)/(\|d_1\|.\|d_2\|)$, where $\bullet$ represents the scalar product, $\|\ \|$ and represents the vector length.

6.4.7 Visualization of Results

After the cluster analysis this module composes reports to allow visualization, understanding and evaluation of the results. The documents are generated in html format and comprise: report, table, table with links, graph and map.

The documents' report supplies some details about the modules execution parameters and a summary of its results. Table documents are used in the exhibition of data lists, in alphabetical or numerical order. Tables with links are similar to the Table, but they add links to the interrelated documents. The option of graphs presents different graphs of the GA behavior.

The map pages use clickable image maps, representing, through a scale of colors, the relative distance of documents inside one group in relation to the centroid of the group and the document that is faraway from the centroid. The links of the clickable image map allow loading grouped documents and the alternate texts of the links present the most frequent terms in the ordered vectors–doc, making possible some comparison between documents before loading them.

6.5 Experimental Results

In the carried tests we used documents obtained from a search in the Google site [6].

The documents are written in English, and they composed four sets of one hundred documents, enclosing the keywords: AIDS, cancer, Karl Marx and Artificial Neural Networks (ANN). These documents generated the 5 bases of tests, one for each subject and another one with all documents.

The tests aimed to evaluate the coherence of the formed clusters and not its precision, because the ideal groupings for the considered bases of tests were not known.

The following parameters were used:

- elimination of radical emptiness (stop words without suffixes);
- elimination of suffixes;
- arrangement *tf*idf*;
- one term for each ordered vector–doc;
- normalization standard *tf*idf*;
- elitism;

- 200 generations;
- initial population being composed of 20 chromosomes;
- cosine metric and super population.

Next, we present the results from the application of SAGH System over the five sets.

6.5.1 Base of Test: AIDS

In the test made with the 100 documents about AIDS, 21 groups were found, being 14 of them with only one document. Since one individual do not apply as a cluster, they were discarded. The other 2 groups had 2 documents and were not considered.

Table 6.1 shows the five best results of this group of documents.

Table 6.1. Best results from the AIDS base

Group number	Number of documents	%	Main Terms
1	23	23	person, doctor, post, posit, thailand
2	14	14	brazil, patent, brazilian, chequer, dentzer
3	13	13	cell, vaccin, seroneg, dose, infant
4	17	17	educ, sector, action, botswana, africa
5	15	15	homosexu, econom, ngo, myth, arv

Group 1 was formed by documents on HIV virus, relating details of the epidemic, forms of transmission, and the tests for virus identification (term "posit").

Group 2 included documents about the Brazilian program of AIDS treatment and prevention, focusing on the break of patents.

Group 4 joined documents that have in common the educational aspect in the prevention of AIDS and its importance to diminish the discrimination. Some documents discuss the absence of educational politics as justification for the enormous expansion of the AIDS in Africa.

The results for this base of test can be considered very consistent, mainly, in relation to the group 2, a group formed by very similar documents.

6.5.2 Base of Test: Cancer

For the test made with documents about cancer, 21 groups were found, but 16 of them contained only one document and 1 contained two documents. The most representative clusters are presented in Table 6.2.

Table 6.2. Best results from the cancer base

Group number	Number of documents	%	Main Terms
1	17	17	prostat, osteosarcoma, psa, ovarian, unknown
2	28	28	bone, tumor, anemia,colon, white
3	26	26	fatigu, cancerbacup, agenc, info, bladder
14	12	12	breast, mastectomi, implant, hung, esr

In this test, the group 1 was formed by documents treating or citing the cancer of prostate, also having some documents about the ovarian cancer and the osteosarcoma. The osteosarcoma documents would have to be on group 2 that has "bone" as one of the terms.

Group 2 gathered documents that approach or refer to the "bone" cancer, being 15 in the total. The term "bone" is also present in other 5 documents, but they discuss other themes. The group number 14 was the most perfect group; therefore it contained, exclusively, documents that dealt with breast cancer.

The results of this experiment can also be considered very coherent, mainly, in relation to group number 14.

6.5.3 Base of Test: Karl Marx

In the tests with Karl Marx as subject, we did not find good results. So, we had to increase the chromosomes number to find clusters with a higher degree of similarities. Then, we found 25 groups, but 19 among them had only one document.

The results are described in Table 6.3.

Analyzing this table, we perceive that the terms "manifesto" and "leagu" appear as main terms in more than one group, evidencing the difficulty in separating documents.

Group 5 was the only one that was formed by documents that dealt with Marxism. The group number 6 had documents that explore the globalization and its relationship with the Marx and Weber theories.

According to the previous experiments, these results were very appropriate, with similar documents inside the clusters.

6.5.4 Base of Test: Artificial Neural Network

In the test with documents about ANN, we did not have a significant division of documents in the first execution. So we applied the same strategy of increasing the number of chromosomes, as carried through in section 3.3. After that, we

Table 6.3. Best results from the Karl Marx base

Group number	Number of documents	%	Main Terms
1	19	19	labour, manifesto, feuerbach, lassal, hobsbawm
2	9	9	soviet, leagu, pipe, truman, shannon
3	8	8	hegelian, travel, shortli, leagu, communitarian
4	25	25	proletariat, bourgeoi, bourgeoisi, manifesto, properti
5	9	9	labor, racism, robinson, doctrin, encyclopedia
6	11	11	global, weber, phil, frank, illuminati

found 25 clusters, where 17 among them contained only one document and 2 groups had 2 documents.

The main clusters are listed in Table 6.4.

Table 6.4. Best results from the Artificial Neural Network base

Group number	Number of documents	%	Main Terms
1	12	12	document, cluster, popul, option, websom
2	19	19	rbf, hidden, element, mlp, propag
3	11	11	art, net, prototyp, cluster, brain
5	12	12	cnn, delai, cellular, precipit, cpg
6	10	10	brain, prospect, connectionist, verb, symbol
7	15	15	ann, hopfield, fire, axon, dendrit

The presence of the term "cluster" in more than one group explains the need to make additional executions.

In this experiment the Group 1 was formed predominantly by documents about the use of neural networks in data mining applications. Group 3 included documents about ART networks. Group 7 added, mainly, documents that introduce the subject of ANN.

The results of these tests involved groups not as homogeneous as found in the previous tests, but we observed that there is a predominance of similar documents in the formed clusters.

6.5.5 Base of Test with all Hypertext Documents

For the base enclosing all 400 documents, 4 groups of 100 documents were found accurately. Table 6.5 shows these results.

Analyzing the main terms of each group, we perceive that they represent the 4 groups in a perfect way.

Table 6.5. Results from the four mixed groups

Group number	Number of documents	%	Main Terms
1	100	25	hiv, aid, infect, brazil, epidem
2	100	25	cancer, tumor, breast, bone, prostat
3	100	25	marx, marxism, capit, revolut, communist
4	100	25	network, neural, input, layer, neuron

6.6 Conclusions

In this work we proposed, implemented and tested a Genetic System for cluster analysis for hypertext documents, called SAGH.

The main contribution of this work was to develop a system for hypertext document clustering, exploring a genetic technique that had been previously developed for grouping numerical data [10].

We consider that the experimental results evidenced that the SAGH System is able to cluster similar documents. Even in sets of documents with bigger interrelationship, the used technique identified homogeneous groups, demonstrating to be a new option to make documents clustering.

The test with all documents confirmed that the system is skillful in identifying main subjects, disrespecting the subdivisions of the contexts.

The visualization of results enhances the comprehension of contents, fastening the search of desired information.

The SAGH creates clusters within a reasonable amount of time. It took less than two minutes to present the visualization of results for the 400 documents database and about 30 seconds for the 100 documents.

Suggestions for future works involve the transformation of the system in an application to be used in information recovery, making it possible to be incorporated in sites of search with analysis of clusters, as some already available, such as Wisenut [19]. We can also incorporate some possibilities to explore this system in other languages distinct from English.

References

1. He, Z., Xu, X., Deng, S.: Clustering Mixed Numeric and Categorical Data: A Cluster ensemble approach, Technical report – Harvard (2005), http://arxiv.org/abs/cs.AI/0509011 (Visited on January 2007)
2. Gulli, A., Signorini, A.: Building an open source meta search engine. In: World Wide Web Conference, Chiba, Japan, pp 1004–1005 (2005)
3. Ferragina, P., Gulli, A.: The anatomy of a hierarchical clustering engine for web–page, news and book snippets. In: 4th IEEE International Conference on Data Mining, Brighton, UK, pp. 395–398 (2004)
4. Lee, U., Liu, Z., Cho, J.: Automatic identification of user goals in web search, Technical report, UCLA Computer Science (2004)
5. Zeng, H., He, Q., Chen, Z., Ma, W., Ma, J.: Learning to cluster web search results. In: Proceedings of ACMSIGIR – ACM Conference on Research and Development in Information Retrieval, Sheffield, UK, pp. 210–217 (2004)
6. Google, http://www.google.com
7. de Luca, E.W., Nürnberger, A.: Using Clustering Methods to Improve Ontology–Based Query Term Disambiguation. International Journal of Intelligent Systems 21(7), 693–709 (2006)
8. Soules, C.A.N.: Using Context to Assist in Personal File Retrieval, Ph.D. Thesis, Carnegie Mellon University School of Computer Science (2006)
9. Ferragina, P., Gulli, A.: A personalized search engine based on web–snippet hierarchical clustering. In: World Wide Web Conference, Chiba, Japan, pp. 801–810 (2005)
10. Hruschka, E.R., Ebecken, N.F.F.: A genetic algorithm for cluster analysis, Intelligent Data Analysis 7(1), 15–25 (2003)
11. de Carvalho, L.A.V.: Datamining: a mineração de dados no marketing, medicina, economia, engenharia e administração. 1.ed, Érica, São Paulo (in Portuguese) (2001)
12. Cole, R.M.: Clustering With Genetic Algorithms. Msc. Thesis, University of Western Australia (1998)
13. Han, J., Kamber, M.: Data Mining: Concepts and Techniques, 1st edn. Academic Press, New York (2001)
14. Lopes, M.C.S.: Mineração de Dados Textuais Utilizando Técnicas de Clustering para o Idioma Português. Dsc. Thesis, Universidade Federal do Rio de Janeiro, Rio de Janeiro (in Portuguese) (2004)
15. Lacerda, E.G.M., de Carvalho, A.C.P.L.F.: Introdução aos Algoritmos Genéticos. In: Anais do XIX Congresso Nacional da Sociedade Brasileira de Computação, Rio de Janeiro, vol. 2, pp.51–126 (1999)
16. Porter, M.: The Porter Stemming Algorithm (1980), http://www.tartarus.org/~martin/PorterStemmer (Visited on September 2006)
17. Goldberg, D.: Genetic Algorithms in Search, Optimization, and Machine Learning. Addison–Wesley, New York (1989)
18. di Carlantonio, L.M.: Novas Metodologias para Clusterização de Dados. Msc. Thesis, Universidade Federal do Rio de Janeiro, Rio de Janeiro (in Portuguese) (2001)
19. WISENUT, http://www.wisenut.com (Visited on January 2007)

Author Index

MIX
Papier aus verantwortungsvollen Quellen
Paper from responsible sources
FSC® C105338

If you have any concerns about our products,
you can contact us on
ProductSafety@springernature.com

In case Publisher is established outside the EU,
the EU authorized representative is:
Springer Nature Customer Service Center GmbH
Europaplatz 3, 69115 Heidelberg, Germany

Printed by Libri Plureos GmbH
in Hamburg, Germany